The Geology of Colorado

Illustrated

The Cover

The two cover photos include the Lyons Formation and some Mesozoic sedimentary rocks that were deposited a few hundred million years ago near Colorado Springs. After the sediments were cemented into hard rocks, the Precambrian core of the Rocky Mountains punched upward and raised those flat layers of rock into these thin ribs of sandstone seen at Garden of the Gods. The thin layers of hard sandstone were interbedded with softer shaly sediments, which have been removed by erosion.

These two color photographs are among the 13,000 color transparencies taken by the author over a span of 40 years. Using color for the cover illustration is, perhaps, unfair advertising for a book that contains only black and white illustrations. But, after some soul-searching, the author decided to get even with all of the "bait-and-switch" advertising gimmicks used by car salesmen, insurance peddlers and appliance stores. It is hoped that a few buyers will be attracted to the spectacular, colorful Garden of the Gods on the cover and browse through the book and buy it on its own merits.

Using color on a single plate for the cover added about fifty cents to the cost of the book. With over 300 photographs, the decision was made to keep the book as inexpensive as possible for the college students. Hopefully the reader will be impressed enough by the black and white illustrations to go visit some of the colorful sites in person.

The Author

Dell R. Foutz attended Weber State and Brigham Young University in Utah and earned a Ph.D. at Washington State University in 1966. He was a pilot in the U.S. Air Force and flew with the Army National Guard at Salt Lake City while he was in graduate school. He maintains that "flying low and slow" is the best way to look at surface geology.

After seven years as an exploration and production geologist with Exxon, in Texas, he moved to Grand Junction for a teaching position at Mesa State College. After 22 years at Mesa State, including a few years as department chairman, Dr. Foutz started a partial retirement in 1993.

His varied professional career included parts of six years with the Utah State Engineer's Office as a hydrographic engineer. He worked with Columbia Iron Mines, examining coal deposits in Colorado in 1956. Shell Oil Company used him for a summer in the Wyoming overthrust belt, and he has spent summers with the Bureau of Reclamation (damsite geology) and Western Engineers of Grand Junction.

The rocks in Colorado force a geologist to look at a number of different things, and in 1982 Dr. Foutz wrote a small booklet titled "Where is the Gold on the Colorado River." In the oil shale "boom" he was geologist and a technical writer for the Paraho Oil Shale Company. His consulting projects have included placer gold, sand and gravel deposits, uranium, oil and gas plus the disposal of a 5-million cubic yard pile of radioactive waste at Grand Junction.

Who

This text is intended for two readers; the first is the freshman-level college student getting general education credit in the physical sciences. The second reader is the tourist (or stray geologist) looking for information about some specific area of the state. Please remember that Colorado has enough variety of geologic features that another book as big as this one could be written about items not mentioned in this book

What, Where, and When...

If you are a college student using this book for a credit class, you need to read with your mind always planning ahead for the next test. As you see the information in the book, think about the three variables that the teacher will likely ask for in a test. WHAT sort of geologic thing is it? WHERE is the thing? And, WHEN, in geologic time, was the thing formed.

For example, suppose you are reading about the Maroon Bells. These are some colorful peaks near Aspen, Colorado, and are frequently shown on expensive calendars. It is not enough to know that the Maroon Bells are pretty mountains in Colorado. From a course using this book, you should learn that the Maroon Bells are unusually high, over 4200 meters (14,000 feet, "fourteeners") above sea level and are composed of red, arkosic sediments of the Maroon Formation of Pennsylvanian and Permian ages. They are located in the Sawatch Range, near Aspen, Colorado. If you had a location map with several circles marked on the map, you should be able to select the circle that would include the Maroon Peaks. By now you should feel a strong urge to learn the time scale and the appropriate vocabulary (words such as "arkosic"). That strong urge is justified. You must know the time scale backwards and forward. If you do not learn the time scale (and the vocabulary) you will be like a math student who chooses not to learn how to count. Of course, you will not be expected to learn the name of every peak in the state, but the Maroon Bells are rather unique and are the key exposure of some red rocks known as the Maroon Formation.

ACKNOWLEDGEMENTS

The writer wishes to thank the Mesa State College administration for a semester of sabbatical leave to get some of the pictures for this book. Betty Herman, when manager of the college Print Shop, assisted with layout and computers and provided many helpful suggestions with the work. Yvonne Peterson helped with half-tones, and Katie Kaufmanis put the whole book in the same computer format. Re-paging was done by Debbie Phipps, and a few special photos were provided by Jack Berry and Michael Eatough. Shari Reeder patiently solved for me a number of frustrating computer problems, and the staff at Pyramid Printing of Grand Junction was especially helpful in all stages of the book and its pre-publication ancestors.

Special thanks go to my wife, Sherry, for waiting patiently while I scrambled up a hillside or leaned over a cliff for a better photo. In her own words she "can't imagine why we should drive 50 miles out of the way just to look at a rock!" Without her support this project would never have been attempted.

CONTENTS

Introduction: The Basics

Geologic Time Scale

The expression "once upon a time, long, long ago..." is ideal for a child's bedtime story. And the expression 704 x 10^9 years is a splendid date for a geophysicist, but, for this book lets go to the geologic time scale. The chart is precise enough for general geology and it also reveals a little of the intrigue and romance of the subject.

In the early years of the science of geology some fossils were found in a sequence of sedimentary rocks in Wales. Many specimens were collected and some rock hounds with training in biology began to name these fossils using the rules of biology. The rocks were described in some detail and then other researchers found the same fossils in similar rocks a long way away—even on the European continent. In order to distinguish these fossils, first found in Wales, from totally different fossils from totally different rocks, they called them "Cambrian" after the name Cambria, which the Roman conquerors gave to Wales. The rock sequence in Cambria became the "TYPE SECTION" for rocks with that age fossils.

Devonshire (England) had very different sediments which were found to overlie the Cambrian rocks. These "Devonian" rocks had unique fossils too, so the outcrops in Devonshire became the type section for the "DEVONIAN PERIOD" of geologic time. Today in Colorado when we find fossils that are similar to those found at Devonshire, we know the rocks that contain them were deposited in the Devonian Period.

Other sedimentary rocks were found in England sandwiched between typical Cambrian rocks and typical Devonian rocks and the scale was expanded to include ORDOVICIAN and SILURIAN rocks.

Of course every good and simple system eventually gets messed up when more facts are known. The geologic time scale has evolved. Note that ERAS are very long spans of time. They are separated by profound worldwide episodes of turmoil in the earth's crust. A worldwide turmoil is known as an OROGENY.

The very long eras are subdivided into formal geologic PERIODS and the periods are divided into formal EPOCHS of time. Working geologists have subdivided all the periods and even the Precambrian Era, but for the

ERA	PERIOD	YRS AGO	EPOCH
	Quarternary		Pleistocene
		2 million	
CENOZOIC			
			Pliocene
	Tertiary		Miocene
			Oligocene
			Eocene
			Paleocene
		65 million	
	Cretaceous		
MESOZOIC	Jurassic		
	Triassic		
		248 million	
	Permian		
	Pennsylvanian		
	Mississippian		
PALEOZOIC	Devonian		
	Silurian		
	Ordovician		
	Cambrian		
		590 million	
PRECAMBRIAN			
		4.5 billion (crust first hardened)	

To help remember the periods, when starting from the bottom, Cambrian, Ordovician, etc., if the first letter is used the sequence becomes ***C O S D M P P T J C T Q****. These become: Creation of Strong Dinosaurs Motivated Puny Primates To Jog Cretaceous Trails Quietly!*

needs of this book, only the Cenozoic Era will be split down to epochs, and the Precambrian will stand as an undivided era.

Each formal period of geologic time was applied to a specific set of rocks somewhere in the world. The English very aggressively published their "type sections" in the early literature, so many type sections are in England. The Permian, however, is named from outcrops in Russia; the Jurassic from the Jura Mountains of Europe and the epochs of the Italian "Tertiary" were established by an Englishman using deposits on an island in the Atlantic Ocean.

In most cases the sequence of rocks used for the type sections has a local disturbance that provides a convenient boundary for the period. However, the disturbance that makes a good boundary between the Pennsylvanian and Permian time in Europe is not found in Colorado, so several rock units in Colorado end up as "Penn-Permian" in age. The basal parts are definitely Pennsylvanian but the top is definitely Permian. In fact, the Europeans have the CARBONIFEROUS Period instead of the North American Mississippian and Pennsylvanian Periods. Both U.S. periods are appropriately ignored by the rest of the world.

Relative Age vs. Absolute Age

In the early days the actual age of Cambrian or Jurassic was "RELATIVE AGE." That means the Jurassic rocks are always younger than Cambrian rocks because they were deposited somewhere above them in sequence. Early workers guessed that the Jurassic was a great deal younger because seven other periods came and went between Cambrian time and Jurassic time. Many attempts were made to calculate the "ABSOLUTE AGE," in years, for the periods. As an example, geologists measured the thickest Cambrian sequences they could find. Then they calculated how long it takes for the same amount of limestone, sandstones or muds to accumulate in modern sedimentary basins. If they could measure 8 mm of lime ooze on a 100-year old shipwreck in a quiet lagoon, they guessed that 80 mm of lime ooze in a quiet lagoon would take about 1000 years. If a rock section had 8000 mm (8 meters or 24 feet) of lime ooze then the time required to deposit that ooze would be on the order of 100,000 years. Today we find limestone sequences thousands of feet thick in virtually every period of geologic time. A wise geologist avoids gravel deposits and coarse sandstone, or volcanic ejecta, because these high energy materials can accumulate in tens or hundreds of meters in a few hours. During the 19th Century the calculations continued and the age of the rock record got older and older. Hundreds of millions of years of fine-grained sediments and limestones were available. Nobody seriously tried to decipher the Precambrian with its metamorphosed sediments. The age was vast!

Then after about 1930 the ABSOLUTE AGE "clock" was perfected. This clock is based on the radioactive decay of several natural elements. A radioactive variety (ISOTOPE) of the element rubidium is Rb^{87}. As this variety (isotope) of rubidium decays it disintegrates to form an isotope of strontium (Sr^{87}). All Rb^{87} is radioactive and Sr^{87} occurs only as the decay product of Rb^{87}. Scientists have determined that half of the Rb^{87} in a rock will decay in 47 billion years. Half of what remains will be gone in another 47 billion years. Therefore, 47 billion years is termed the HALF LIFE of Rb^{87}.

This isotope of rubidium occurs in molten rocks. The

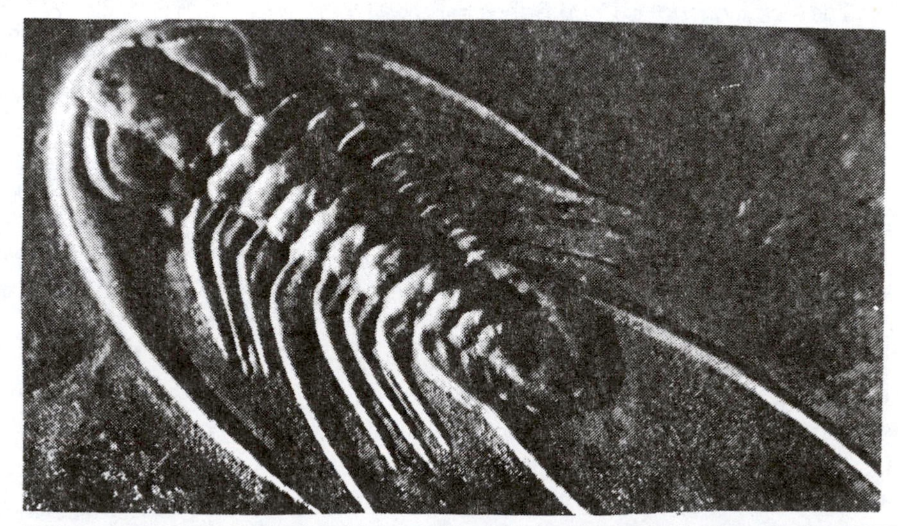

Fig. 1. Trilobites dominated the seas during the Cambrian Period. This is one of the many distinctive genera of this extinct arthropod. A variety of species characterize early, middle and late Cambrian marine sediments. They were abundant only in early Paleozoic seas. The last one died in the Permian.

rubidium is locked into crystals as the melt freezes. When a crystal captures an atom of Rb^{87}, the clock is set. A kilogram of igneous rock may contain millions of atoms of Rb^{87}, which will decay at the standard rate. Half of the atoms will decay to Sr^{87} in 47 billion years. It may sound incredible, but the system is remarkably accurate, no matter what conditions the rock sample experiences.

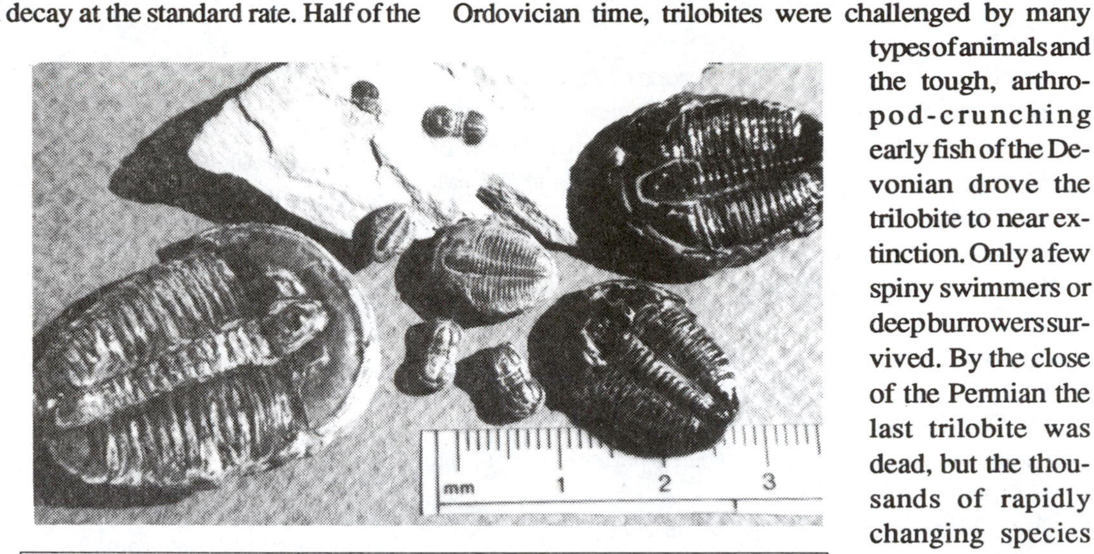

Fig. 2: This group of trilobite fossils was recovered from a shale deposit in western Utah. These occur only in middle Cambrian sediments.

Today's laboratories can count the ratio of Rb^{87} to Sr^{87} in a sample in a few hours and give an age for the time of crystallization within a few percent for about $400 per test. The dates are not limited to large masses of rock or thick lava flows either. Thin ash layers from erupting volcanoes are ideal page markers in our book of sedimentary rocks. Ash can spread hundreds of miles from the volcano and usually can be recognized and dated in a sequence of sediments. The relative dates that the fossils provided now have been tied to rather accurate absolute dates in all periods.

In addition to Rb-Sr dates, there are many other decay series. Several isotopes of uranium decay to unique isotopes of lead. Each pair has a different half-life. $Potassium^{40}$ is radioactive and decays to $argon^{40}$. The decay pair which is used on very young geologic features is $carbon^{14}$, which decays to $nitrogen^{14}$ in a half-life of only about 6000 years, and is effective for determining ages back to about 40,000 years ago. When a glacier or volcanic cloud or lava flow buries a tree branch, and the age is less than 40,000 years, it can be dated quite accurately. Archaeologists rely heavily on C^{14} dates.

The fossil record is far more accurate than most people realize. Trilobites were the only really abundant marine critter of the Cambrian seas. For over 60 million years this unique group of arthropods ruled the world, and no multi-celled critters lived on the land. By Ordovician time, trilobites were challenged by many types of animals and the tough, arthropod-crunching early fish of the Devonian drove the trilobite to near extinction. Only a few spiny swimmers or deep burrowers survived. By the close of the Permian the last trilobite was dead, but the thousands of rapidly changing species had left a profuse record of trilobite fossils in thick piles of marine sediments around the world.

Fig 3: This specimen of <u>Asaphiscus</u> appears as the middle-Cambrian shale material is stripped away. The specimen is about 4 cm in length.

Paleontologists monitor the rock cuttings that are pumped to the surface of oil wells as they are drilled from the surface to the deeply buried reservoir rocks. In one area of Texas, a little marine, bottom-dwelling single-celled microfossil named *Textularia warreni* is the key to a dangerous high pressure gas zone in the Austin Chalk. When "*Tex warreni* variety 18" appears in the cuttings, it is time to stop drilling and add steel casing to the hole and drill ahead carefully into the gas zone.

Only a few feet of rock record the rise and extinction of old "*Tex warreni* 18." If you found him, you built a safe oil well. If you didn't recognize him, the well could blow wild with catastrophic loss of equipment, petroleum resources, and even lives. Precise fossil ages can make the difference!

Mass Extinctions

Although mostly beyond the scope of this book, there are some fascinating moments in geologic time that bear mention. At the close of the Permian and again at the end of the Cretaceous, nearly all life, worldwide, on land and sea, was snuffed out.

The number 90% is appropriate at the close of the Permian. Not just 90% of all the specimens of a particular critter, but 90% of all *kinds* of life. Virtually all genera, nearly every taxonomic family and even some orders of biology became extinct. It was truly a MASS EXTINCTION. Climate change, volcanoes, comet tails and meteorite collisions are suspects in the losses, or perhaps a combination of them.

At the end of the Cretaceous the extinctions were not quite so thorough, but that boundary is much easier to study, and geologists and biologists are scouring the Cretaceous-Tertiary boundary for any shred of evidence that may answer the question of how such a disaster may have occurred.

Another interesting puzzle in the development of the fossil record is the nearly EXPLOSIVE APPEARANCE, and rapid development, worldwide, of hard-shelled marine animals at the beginning of the Cambrian. The earliest one-celled organisms—like plants and/or bacteria—have been found in rocks about 3 billion years old. But somehow the air, water, life and land waited until the very end of the Precambrian before the real exciting fossil story begins in earnest. All major marine fossils had their earliest ancestors in the Cambrian, although most groups struggled until Ordovician time to become really abundant. The individual animals were totally unlike today's sea life, but the profusion of marine life in Ordovician time was like our modern oceans. Land plants waited until Devonian time to make their mark. Again, the individual plants were unlike today's plants, but the profusion of greenery was similar to our present world (even though the continents had funny shapes and were in strange places).

Summary of Colorado's Geologic History

Before we drown in the details of the geology of Colorado, a brief overview can put the major events into perspective. Details can be stuffed in later.

Precambrian: Vast basins received marine and continental sediments, many kilometers thick, which were buried, metamorphosed and injected with igneous rocks of all sorts. (There, that is the Precambrian in seven seconds!) These altered sedimentary rocks became the gneisses, schists, quartzites, slate, marble and granite of the so-called BASEMENT ROCKS of all the areas of the state. These old and relatively hard rocks are exposed in the core of many of the mountain ranges of Colorado where the younger rocks, if they were present, have been stripped off by erosion.

Cambrian: An ocean advanced from the west to cover a lifeless landscape. After crossing Utah, the advancing sea began to attack Colorado in middle Cambrian time. As the platform subsided and/or the waves subdued the mountains, the resulting ground-up mountains became the beach sands of the Sawatch Sandstone. In the northwest portion of the state, the sands are a bit redder and they are called the Lodore Sandstone.

The beach environment migrated to the east, beyond the present Front Range, and in late Cambrian time the Peerless limes accumulated in the plains area. At the end of the Cambrian Period, only a 100 X 300 km. wedge of dry land remained in Colorado. Marine limes that filled a narrow trough from Eagle to Utah became the Dotsero Limestone. Maximum Cambrian sedimentation is along the Utah boundary, attaining a thickness of about 270 meters (900 ft.), which is mostly sandstone. A minor episode of unique mineralization, including gold, was emplaced at Powderhorn, southwest of Gunnison.

Ordovician: Most of Colorado was elevated during Ordovician time, either not receiving sediments, or having them removed by subsequent erosion. A small area in the Glenwood Springs area received some magnesium-rich limestone, called DOLOMITE, in what is called the Manitou Formation. The unit is named from exposures at Manitou Springs, and the rocks form some of the features in nearby Cave of the Winds. Additional carbonates are deeply buried under the Kansas-Oklahoma corner of the state. Marine sediments are rather thick both east and west of Colorado, but during Ordovician time, most of the state was a part of a broad arch and remained dry.

Silurian: The broad arch of Ordovician time continued and *no Silurian sedimentary rocks are recognized in Colorado*. If deposits occurred, they were swept away by erosion. There are a few Silurian (?) diamond-bearing, igneous pipes (cylinders) near the Wyoming border in the Front Range, but do not invest your life savings to find a fortune in diamonds in Colorado, as the "stones" are tough for the experts to recognize because of the small size (average .01 carat)!

Devonian: Sediments of Devonian age are limited to about 200 meters of Ouray and Elbert limestones in the southwest corner of the state, plus a thinner sequence of sand and limestone (Parting and Dyer Formations) in a strip that includes Glenwood Canyon. Dyer equivalents (limestone) spread eastward from near Colorado Springs to underlie the entire Kansas border.

Mississippian: The west half and the southeast corner of the state had significant marine embayments during this period which left up to 200 meters of fossiliferous Mississippian limestone. Much of the limestone in the west was later removed by erosion, but the remnants provide a host rock (the Leadville Limestone) for some prolific deposits of lead, zinc, gold, silver and other valuable ores.

Pennsylvanian: This period marked great changes in Colorado. Two great islands, "UNCOMPAHGRIA" and "FRONT RANGIA" surged upward and the debris shed from them filled the adjacent basins with up to 3500 meters (over 10,000 ft.) of red, sandy sediments that provide much of today's mountain scenery for the region. The Paradox Basin, in the southwestern corner, and the Maroon Basin in the Eagle area became isolated from the sea and received great volumes of salt and gypsum. These pliable sediments later were squeezed around to make some unusual salt domes. While the big islands were present, thick aprons of gravel and red, continental sediments accumulated around them. The Maroon Formation filled the center basin, the Cutler Group filled the southwestern area, and the Fountain Formation was dumped on the east slope of the Front Range. These thick layers are vividly exposed now in the 4300 m. (14,000 ft.) Maroon Bells, the colorful 4-Corners

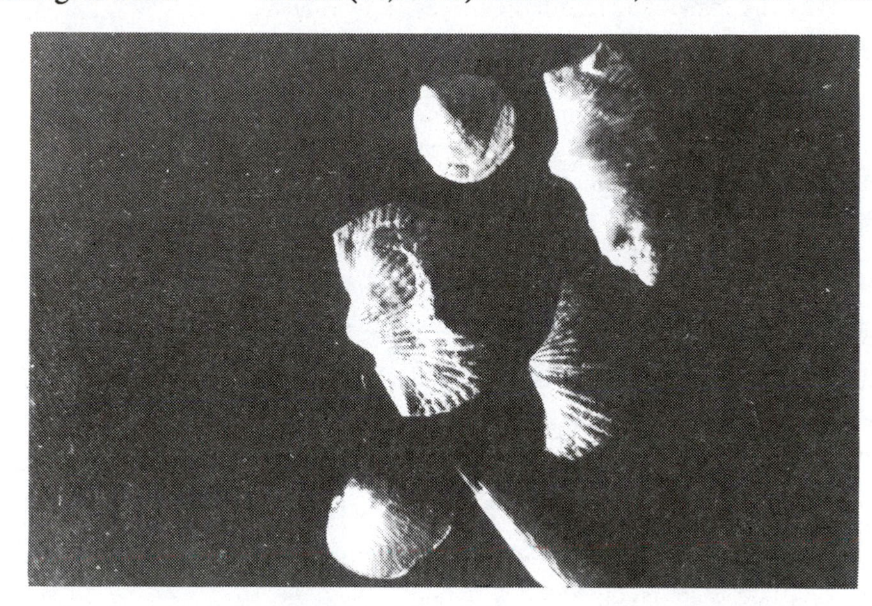

Fig.4. Before the two big Pennsylvanian islands popped up in Colorado, a variety of marine animals lived in the ocean of the day. These are four different, extinct brachiopods, or lampshells. A pencil point gives scale.

Fig. 5. This productid brachiopod (lampshell) lived near Glenwood Springs during the early Pennsylvanian, when a tropical sea covered the area. A pencil point marks the scale.

scenery and some of the impressive Front Range features from Garden of the Gods and Red Rocks Park to the Boulder Flatirons.

Permian: In most of Colorado, the Permian Period is a continuation of the events of the Pennsylvanian. Just west of the state, the Permian featured a shallow sea, teeming with biology, stretching from Mexico to Alaska, and Colorado was host to two of its great islands.

World Crisis: To close the Permian, a worldwide catastrophe ended the Paleozoic Era when the American continents were ripped away from the world's supercontinent ("Gondwanaland") and the Atlantic Ocean was born. The American plates began rafting their way into the Pacific basin, and the great seaway west of Colorado abruptly became a barren plain, roamed by an entire new biology, dominated by terrible lizards called dinosaurs. The Permian was not very impressive in Colorado, but the end of it was unbelievable, and on a global scale.

Triassic: The Mesozoic Era starts with red muds, coarse sands and gravel being deposited around the persistent uplands of Uncompahgria and Frontrangia. These two islands merged to occupy the central half of the state. Basins in the SW, NW and SE corners of the state received a few hundred meters of bright red sediments, which are mostly continental. Geologists have learned that marine conditions leave generally gray sediments, but continental environments produce brightly colored (oxidized iron) red and yellow sediments. In Colorado the Triassic and Jurassic rocks are colorful! Colorado National Monument, at Grand Junction, is mostly Triassic "redbeds" lying on ancient, dark metamorphic Precambrian rocks. The Garden of the Gods is chiefly Mesozoic and Paleozoic red beds.

Jurassic: The Morrison Formation is the big feature of the Jurassic, and this colorful, bone-rich sequence of continental shales, stream gravels, sandstones and thin freshwater limestones is recognized over all of Colorado and Wyoming, plus parts of 10 other Western states. And dinosaurs *owned* the world in the Jurassic.

Cretaceous: This is the one period, 80 million years of time, that excites the historical geologist in the West. The 330-page *Geologic Atlas of the Rocky Mountain Region* devotes a full 38 pages to just the sedimentary history of the Cretaceous Period. Another ten pages deal with oil, gas and coal which are important in Cretaceous rocks of the area. Possibly 25% of the geologic story of Colorado, from out-crops to economics, is in the Cretaceous - a scant 2% of the geologic time scale.

In a nutshell, a Cretaceous sea advanced over the state leaving the Dakota Sandstone as a blanket of tideland and beach deposits; which were covered statewide by gray shales of the shallow marine Pierre and Mancos formations. There, that is the Cretaceous of Colorado.

To add a few details, a few large islands lay to the west of the state from Arizona to Montana, never yielding to the vast shallow sea that eventually stretched from the Gulf of Mexico to Alaska. These islands finally surged upward, forming a lofty chain of mountains that shed sediment to the east, filling most of the seaway in Colorado, and the remaining waters withdrew eastward and then to the south to become the present Gulf of Mexico. The advancing deltas and tidelands of the Late Cretaceous included vast, steaming swamplands which were buried under thousands of feet of stream-born sands, thereby preserving about one third of the good coal resources of the United States in a wide strip from Arizona to Montana. Thus ended the last marine occupation of Colorado.

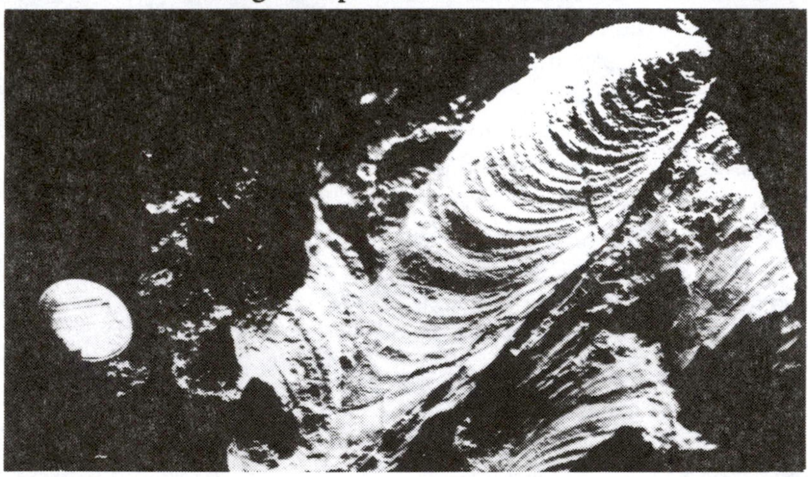

Fig.6. The shallow, somewhat murky ocean of late Cretaceous time teemed with creatures of many biological groups. This Inoceramus clam is rather characteristic of the time, and is exposed near Grand Junction, in the Mancos Shale. A penny shows the scale. Note the snail borehole near the upper right tip, probably indicating what killed this one. All Inoceramus were gone at the close of the Cretaceous.

Laramide Orogeny: The Laramide OROGENY is not a geologic period, but is an upheaval of the earth's crust of catastrophic proportions. The disturbed rock layers were first recognized in the Laramie Range of mountains, which is the northward extension of the Front Range into Wyoming. This disturbance began in late Cretaceous time and

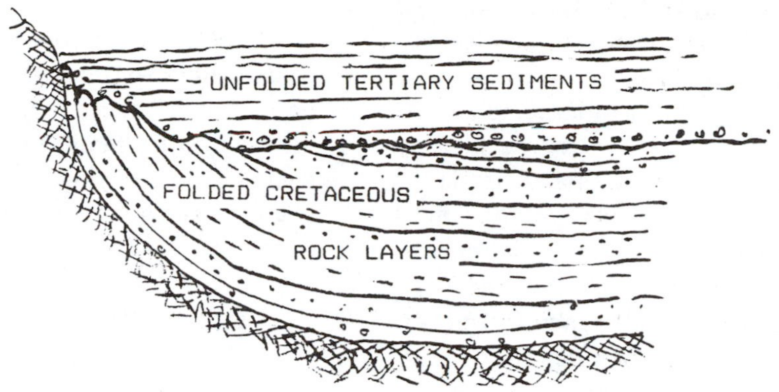

Fig. 7. *The Laramide Orogeny is the mountain-building episode that separates the Mesozoic Era from the Cenozoic Era. Evidence of the turmoil is best revealed just north of Colorado, in the Laramie-Front Range.*

lasted about half way through the Tertiary. The age of the Laramide orogeny was first defined near the Wyoming-Colorado border where the youngest fossil-bearing sediments that are wrenched upward give a rather precise date (late Cretaceous) for the orogeny. This worldwide upheaval reached its climax with the building of a chain of mountains from the Aleutian Islands of Alaska, through North and South America, and reaching to the southern tip of Chile. In Colorado, the disturbance began as a dozen or so small intrusions (STOCKS) poked upward, starting at Ute Mountain, near the 4-Corners. By mid-Tertiary time we had all the Rockies assembled and receiving their last few "shots" of mineral-rich vapors which were crystallizing in their veins. These would become the treasures that would put places like Cripple Creek, Telluride, Silverton, Creede, Leadville, Blackhawk, Fairplay, Golden and Aspen on the map.

Tertiary: After describing events of the Laramide Orogeny there is not much to say about the Tertiary Period. It was the time to finish building the mountains and to decorate them with outpourings of lava and piles of volcanic ash that add variety to the state. The mountains enclosed a 3-pronged basin with centers in western Colorado and adjacent parts of Utah and Wyoming; and the basin filled with fresh water and hundreds of feet of unique sediments that enclose, today, the greatest sequence of oil shale in the world. Indeed, all the oil of the fabulous Middle East could be matched barrel for barrel by oil from the shales of this relatively small area. But the oil would be expensive to extract.

Coal was deposited from extensive Western swamps, including the Denver area, as well as much larger deposits in the states to the north. North Dakota leads the nation in coal supplies, but the coal is LIGNITE (very low grade coal) and has limited commercial use. Wyoming and Montana also have great reserves of low grade Tertiary coal.

Near the end of the Tertiary the entire Southwest U.S. was broadly uplifted by several kilometers, allowing erosion to cut deeply into the plateaus and mountains. The Grand Canyon of Arizona is one famous canyon that was carved in the late Tertiary, but Glenwood Canyon, Black Canyon of the Gunnison, the Royal Gorge, Flaming Gorge, and others were carved then too. And all the mountains became more rugged as well.

Quaternary: This period is the ICE AGE. Its short two million years put the frosting on the beautiful Colorado cake. The mountain scenery needed the gouging and trim that only glaciers can perform to create the really rugged alpine terrain that characterizes the state. The ice has been gone a short 10,000 years, and the magnificence of Colorado is revealed for us to study.

Minerals

Any discussion about the geology of Colorado requires some basic understanding of rocks, minerals, geologic time and some basic vocabulary. If the reader has no experience with these topics, this section should help.

ROCKS are aggregates of minerals, so lets decide what a mineral is and focus on a dozen basic minerals needed to find our way around the geology of Colorado.

MINERALS are "naturally occurring, inorganic solids with definite physical and chemical properties." The definition is not perfect because "mineral oil," coal and oil are organic and do not fit the definition, yet coal is one of the sedimentary rocks. To a physicist, glass is not solid, yet obsidian (volcanic glass) is the quickly cooled rock produced by some volcanoes. Introductory classes in geology usually include laboratory periods to teach rocks and minerals, but this book is in a hurry to get into the geology of Colorado. We will force-feed a few vital minerals and hope the students have access to displays of the critical minerals to speed their familiarity. Twelve common minerals are essential. Learn their names, origin and a few physical properties for each of them. The physical properties include such things as hardness, color, density, cleavage and luster. Many other physical properties exist, such as radioactivity, taste, flexibility, magnetism and so on, but we will skip on to the needs of this book.

HARDNESS is the resistance to scratch. The hardest natural substance is diamond. We give it a hardness of 10. It can scratch the next hardest index mineral, corundum, which has a hardness of 9. Of course you are not interested in corundum, unless you know it forms the gemstones ruby and sapphire. The index minerals on the HARDNESS SCALE are shown in the box below.

Hardness scale

diamond	10
corundum	9
topaz	8
quartz*	7
feldspar*	6 (orthoclase and plagioclase)
apatite	5 (knife blades and glass are 5.5 and 6)
fluorite	4
calcite*	3 (fingernails are about 2.5)
gypsum*	2
talc	1

(* indicates minerals on the "essential list")

There are about 2000 minerals and each fits somewhere on this scale.

Essential Minerals

1. feldspar (2 kinds)	7. mica (2 kinds)
2. quartz	8. gypsum
3. calcite	9. kaolin (and other clays)
4. limonite	10. gold
5. hematite	11. pyrite
6. amphibole (hornblende)	12. molybdenite

No other geologist would pick the same 12 minerals, especially in this order, but for this book, it is a great list!

CLEAVAGE is the tendency of a mineral or rock to split cleanly along planes of internal weakness. Mica splits easily into thin sheets, exhibiting perfect cleavage in one direction or one "family of planes." The sheets of mica part somewhat like playing cards separate. They may be tough to tear in half, but the individual sheets (cards) slide apart easily.

COLOR, REACTION TO ACID, LUSTER, SPECIFIC GRAVITY and other physical properties help us to distinguish the minerals from one another. The 12 essential minerals for this book are indicated in the lower box.

FELDSPAR: All rocks began as a hard crust on a molten earth, and the dominant mineral that crystallizes from molten rock on this planet is feldspar. If we should melt the earth, we would have a liquid loaded with 8 important elements, and nearly 100 lesser elements. The 8 most abundant elements are:

oxygen (O)	calcium (Ca)
silicon (Si)	sodium (Na)
aluminum (Al)	potassium (K)
iron (Fe)	magnesium (Mg)

The most common elements are oxygen and silicon, and minerals containing these two elements are called SILICATES. Next is aluminum, and the others are listed in order above. As molten earth began to cool, the various elements began to freeze into a variety of crystals. Oxygen, silicon, aluminum, calcium, sodium and potassium naturally build a structure that we recognize today as feldspar. At high temperatures, calcium fits into the structure, but as the temperature decreases, sodium enters the structure. At even cooler temperatures, potassium goes into the structure.. High temperature feldspars are the calcium and sodium feldspars (PLAGIOCLASE). Cooler feldspars, with potassium, are ORTHOCLASE. Rocks derived from molten origins are IGNEOUS ROCKS. This paragraph on feldspars is pretty heavy reading for the non-scientist, but remember, geology students must wade through Shakespeare and Mozart for their general requirements and it is good for them.

For review,

calcium, aluminum, silicon and oxygen
make hot plagioclase

sodium, aluminum, silicon and oxygen
make cooler plagioclase

potassium, aluminum, silicon and oxygen
make orthoclase

To make this book useful for honest college credit in the physical sciences, lets also use the chemical symbols for the elements in the common minerals. Students who have not already been introduced to chemical symbols probably cheated themselves in their preparation for college.

For honesty,

Ca + Al + Si + O	make hot plagioclase
Na + Al + Si + O	make cooler plagioclase
K + Al + Si + O	make orthoclase

(Notice how much simpler the ingredients can be listed if we use the shorthand symbols.)

Other common symbols that may appear in this book are:

Sr	strontium	Au	gold
Cu	copper	H	hydrogen
Rb	rubidium	Ag	silver
C	carbon	Fe	iron
Pb	lead	Mo	molybdenum
S	sulfur	Zn	zinc
Cl	chlorine		

Orthoclase is very similar to plagioclase in appearance. Both have hardness of 6, have a glassy luster, and have good cleavage in two directions that tend to make right angle corners. Note the piece of feldspar shown in Fig. 8. It has a family of flat surfaces which makes the base flat, where it rests on the table. The top of the specimen is also a series of flat surfaces, parallel to the base, defining one direction of cleavage. The flat left and right sides of this piece show the second family of cleavage planes for feldspars. The two "directions" of cleavage intersect at right angles, making the feldspars rather easy to identify. In Figure 8, the top and bottom of the picture show ragged edges, showing that this specimen does not have a third family of cleavage planes that would tend to make cube-shaped fragments.

QUARTZ: When feldspars use all the silicon (Si) and oxygen (O) needed to make their crystals, there is usually a surplus of silicon and oxygen left in the "stew" of molten rock. If the remaining silicon atoms can pick up four oxygens, the mineral that forms is quartz, with a formula of SiO_2. Low temperature, or "cool" igneous rocks usually contain some quartz. Quartz can be the dominant mineral

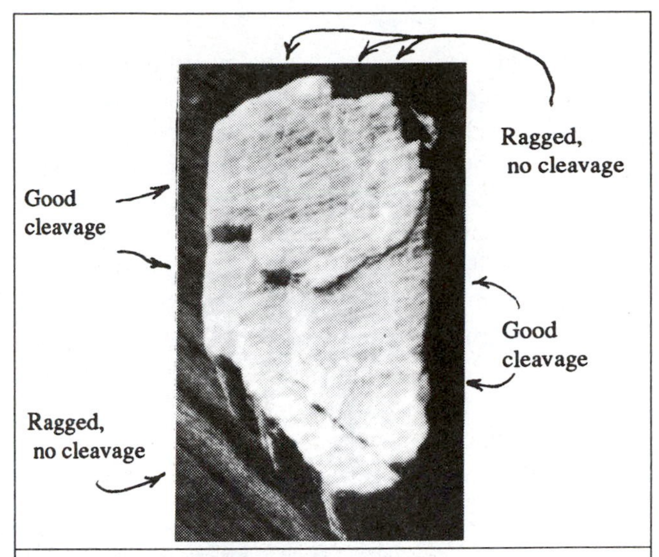

Fig. 8. This is a chunk of feldspar. It is rather pink in color so it is orthoclase. It crystallized at rather cool temperature (cool compared with other molten rock materials). The piece has two directions of cleavage that intersect at right angles.

in some igneous rocks.

Quartz is a relatively hard mineral at H=7, and it has no cleavage, therefore it is hard to break into smaller pieces. It can come in any color, depending on impurities. Quartz has a glassy or "vitreous" luster and, if it has a chance to grow into an open space, it forms a characteristic six-sided prism with a pointed cap on top. When it

Fig. 9. Quartz will form 6-sided prisms with caps on top if it can grow unimpeded into an open cavity. It has no cleavage.

crystallizes from a molten mass, it is among the last to crystallize and it fills the spaces between other crystals and rarely makes prisms. Quartz is the most common mineral in sand. Although feldspars are more abundant in the original crust of the earth, quartz is harder, has no cleavage and is less soluble in the weak acids of the environment, therefore, as the rocks are battered by the elements, quartz survives better than feldspar. Quartz was common in the original crust, so it is very common in the later materials.

CALCITE: This mineral is not formed from molten material. It is precipitated from water (ocean, springs or lakes) and is made from the calcium that is released when feldspars and other igneous rocks are dissolved. The calcium combines with carbon and three oxygens to form calcium carbonate ($CaCO_3$). Don't let the numbers in the mineral formulas bother you, it just seems unnatural to write carbonate without using the correct number of oxygen atoms in the structure to make it "carbonate." This is not a book in chemistry, but most people are familiar with calcium carbonate in the form of limestone. It is the chief ingredient in sea shells, cavern formations and even the scale that forms in hot water pipes and tea kettles.

Fig. 10. Calcite has three directions of cleavage, forming squashed cubes or "rhombs." dilute hydrochloric acid (HCl)will fizz on calcite.

Calcite is relatively soft, with H=3. It may be transparent or almost any color, depending on impurities. If a drop of dilute hydrochloric acid is applied to a carbonate mineral the carbonate will neutralize the acid with a vigorous bubbling reaction (fizz). Calcite has excellent cleavage in three directions that intersect at an angle of 76 degrees. Cleavage fragments of calcite have a distinctive rhombohedral shape, but the limestone and sea shells do not show the cleavage shapes because they contain microscopic calcite crystals. Limestone often includes white veins of calcite crystals which have filled fractures in the brittle limestone.

LIMONITE and HEMATITE: These two oxides of iron are often found together as weathered by-products of iron-bearing minerals. Limonite is hydrous (contains water) and hematite is anhydrous (no water in the structure). Both minerals are familiar as the common rust of iron. Limonite is yellow and brown, and hematite is mostly red. Hematite gets its name from the root "hema" or blood (red). Limonite may be remembered as the yellow ("lemonite") variety of rust. Most of the red colors seen in rocks and minerals result from subtle coatings of hematite. Most of the yellows and browns of rocks and minerals are from varieties of limonite. In fact, most colors in rocks, even the purples, greens, and oranges are from traces of several iron-bearing minerals. The red rocks of the American West are stained with hematite, and limonite is a common yellow stain which may become the clue that more valuable minerals also may be present. Gold, silver, lead, zinc, copper, molybdenum and a host of other metals occur as SULFIDES (metals mixed with sulfur) in ore deposits around the world. These sulfides are extremely rare in the earth's crust compared with iron. Iron is nearly always associated with the more valuable metals, and when a valuable ore deposit begins to weather in nature, the associated iron sulfides rust away with a tell-tale yellow or brown color. Early prospectors quickly learned to prospect for limonite; then they would check to see if the more valuable minerals were also present.

AMPHIBOLE: Amphiboles are a group of black silicates common in igneous and metamorphic rocks. They contain iron and/or magnesium, mixed with Al, Si and oxygen. Calcium, Na and a few other elements may be included, which make for the varieties of the group. HORNBLENDE is a common member of the group, occurring in igneous rocks of intermediate temperatures. Hornblende is black, glassy, and has two good cleavages that intersect at 56 degrees.

MICAS: Potassium-aluminum silicates which can be split into thin, glassy, flexible sheets are the micas. MUSCOVITE is the transparent variety and BIOTITE contains a little more iron and/or magnesium that make it black. Iron and magnesium tend to make minerals black, heavier and of a higher temperature of crystallization. Although micas are common and easy to identify in igneous rocks, they are even more important in the classification of metamorphic rocks.

GYPSUM: When a sea evaporates the dissolved salts that remain may become thick sequences of salt and gypsum. Salt (the mineral HALITE, NaCl) quickly dissolves away in nature, but the associated gypsum often remains as a soft rock at the surface. Thick layers of semi-plastic gypsum and salt (under great overlying pressure) have been squeezed around, much like toothpaste, to form several unique features in Colorado. Gypsum is normally a soft, crumbly white mineral at the surface, but when pure crystals occur, they are transparent and can be scratched with a fingernail. Gypsum is made of calcium, sulfur, oxygen and water ($CaSO_4 + H_2O$).

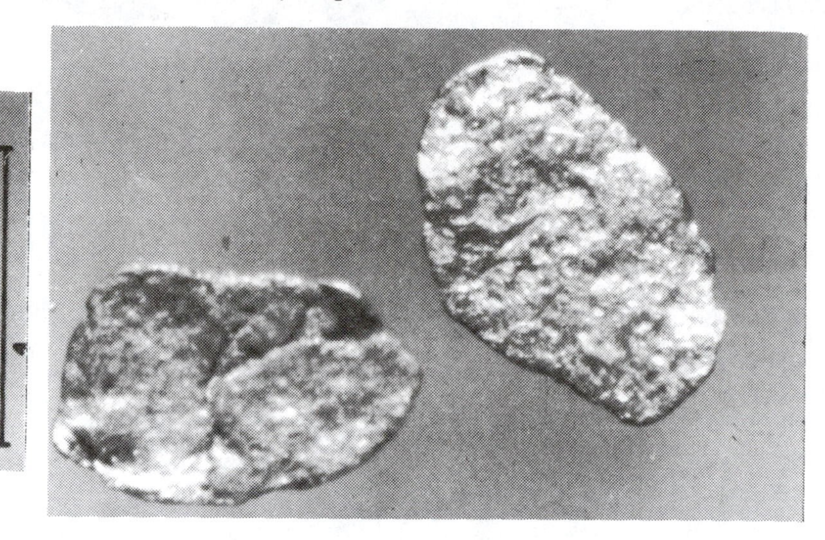

Fig. 11. Two gold flakes recovered from the Colorado River, near Kremmling. They probably entered the stream system near Breckenridge and have been pounded and battered by pebbles and boulders during more than 80 km.of transport. Scale is in millimeters.

KAOLIN: When feldspars weather away the major new mineral formed is kaolin, a soft, flaky clay made of aluminum silicate with included layers of water. If clays are dried excessively, they shrink. When re-wetted, they expand again. Some clays shrink-expand much more than others. Clays form the sedimentary rock known as shale. The most common sedimentary rock is shale.

GOLD: This mineral (the single element, gold or Au) is not listed among the important minerals in most beginning geology books, but this is a book about Colorado! The state is more exciting if we include the early gold rushes in some of the nation's most colorful mining camps. So, for the geology of Colorado, gold is important! It is soft and malleable (easily pounded into new shapes) and so heavy that it weighs 19 times as much as an equal volume of water. Most important - its beautiful metallic luster and resistance to corrosion make it very valuable. Silver is often associated with gold in mineral deposits, and is equally as important in the mining history of the state.

PYRITE: Better known as "fools gold," pyrite is a metallic mineral made of iron and sulfur (iron sulfide, or FeS_2). It has a brassy or golden color and metallic luster, but generally does not behave like gold to an experienced geologist. It weathers to brown limonite and becomes the prospector's best clue in the search for valuable ore deposits. Fresh crystals of pyrite can form as beautiful cubes or 12-sided "pyritohedrons," and when these distinctive crystals weather to limonite rust, the limonite forms a false crystal of the same shape.

Pyrite may form from the cooling vapors of an igneous intrusion. If traces of gold, silver or other valuable metals are present in the molten rock, sulfides of each of these can be associated with the pyrite in veins in and near the igneous mass. Pyrite has a hardness of 6 to 6.5; is brittle and makes a greenish black powder when scratched against a harder substance. The dark powder is significant when comparing pyrite with gold because gold makes only a gold powder.

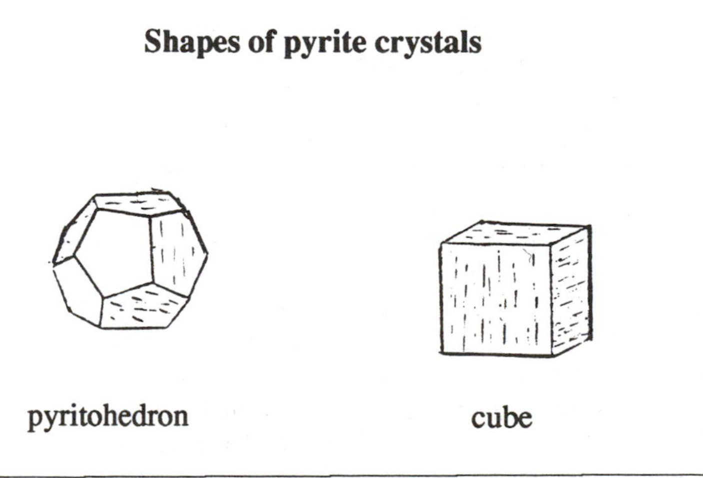

Fig. 12. Brassy pyrite normally forms cubes or pyritohedron crystals. It is brittle and forms a black powder when scratched.

Rocks

Rocks are "aggregates of minerals that must be dug by tools". That is a strange definition, but dirt, soil and loose sediment are not included in our classifaction of rocks. If they are not cemented or stuck together tightly enough to require tools to break, they are not rocks! IGNEOUS rocks crystallize from molten rock (MAGMA). They are characterized by having interlocking crystals randomly arranged. SEDIMENTARY ROCKS are made from the debris of other rocks. They are CLASTIC SEDIMENTARY ROCKS if they are made of particles of former rocks, and they are CHEMICAL SEDIMENTARY ROCKS if they are precipitated from solutions of dissolved rock material. Clastic sedimentary rocks characteristically have packed grains of rock or crystal material cemented together, often with pore spaces between the grains. They are classified on the basis of the grain sizes (sand vs. pebbles) and the composition of the grains (shale is mostly clay minerals). METAMORPHIC ROCKS are made from sedimentary or igneous rocks that have been changed from their original form, but not enough to be remelted as new igneous rocks. They typically have interlocking crystals, but the crystals are aligned in relic sedimentary layers or recrystallized in a new pressure-controlled orientation, with new minerals forming from the original materials, especially clays changing to micas.

Fig. 13. A cliff of conglomerate. Individual clasts are as large as cobbles. Note that horizontal stratification is prominent, a characteristic of all sedimentary rock types.

CLASTIC SEDIMENTARY ROCKS: Clastic sediments are the pieces of rock material that are deposited in a new location. They include the familiar sand, gravel and boulders, but the term SAND is restricted to grains between 1/16 mm. and 2 mm. That seems like a strange and arbitrary size range, but in the world of wind transportation and the suspended grains in a flowing stream, the numbers are important. The classification is based on a logarithmic scale in metric units so be prepared for some strange numbers. The scientific world always uses the metric units because they are the most versatile. If you learn the limits of sand sizes, the rest come easy. SILT grains are those between 1/256 and 1/16 mm. Nobody really measures this classification except under a microscope. When grains are so small that they are hard to distinguish as individual particles with the unaided eye, they are smaller than 1/16 mm. Silt-sized grains feel rough to the touch and are gritty between the teeth. If a sedimentary rock has a gritty surface, even though the individual grains are too small to see clearly, it is likely a siltstone. SHALE is made of particles smaller than 1/256 mm. That number is ridiculous except on a log scale of rock particles. However, it is also the general size of clusters of clay crystals that will settle out of standing water. True shale is made of clay crystals with other typical sand minerals missing. Even quartz, which is nearly insoluble at surface conditions, is slightly soluble in the size range of shale crystals, so it is generally excluded when clean shales are deposited. Often fine sand, silt and shale sizes are mixed in natural environments and geologists lump these mixed sediments as "MUD." CONGLOMERATES are rocks made of particles coarser than 2 mm. Up to 64 mm. (3 inches) they are PEBBLES, and from 64 to 256 mm. (10 inches) they are called COBBLES. Rocks made of particles larger than 256 mm. are called BOULDERS.

Some metrics are a must! A meter is about 39 inches. Now you know the metric system!! The hard part about metrics is converting an excellent system to an archaic, unwieldy system.

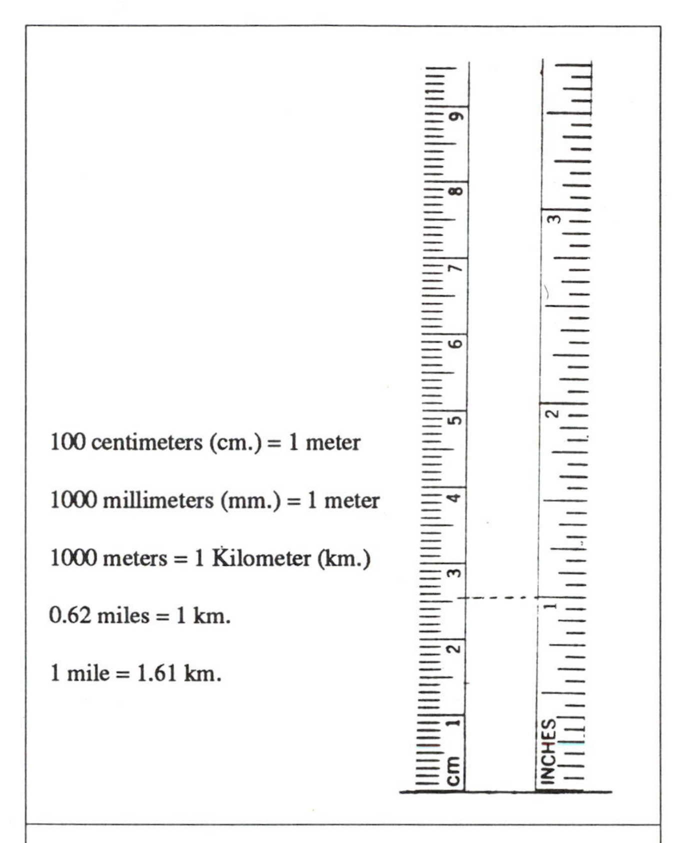

Fig. 14. A comparison of metric and inch scales. Note that 2.54 cm. equal one inch. The centimeters are subdivided into millimeters.

Grain sizes of clastic sedimentary rocks are:

conglomerate

boulder	= grains larger than 256 mm. (10 in.)
cobble	= grains 64-256 mm. (3-10 in.)
pebble	= grains 2-64 mm. (less than 3 in.)

sandstone	- grains $^1/16$ to 2 mm. (quartz + other grains)
siltstone	- grains $^1/256$ to $^1/16$ mm. (quartz + other grains)
shale	- grains less than $^1/156$ mm. (mostly clay crystals)

CHEMICAL SEDIMENTARY ROCKS: A few common rocks precipitate from solution, just like salt will precipitate to form a crust on the bottom of a dish of saltwater as the water is evaporated. LIMESTONE is the most familiar to lay persons. Limestone is made of the mineral calcite. Usually the calcite is extracted from the water by organisms that use calcite to make a hard protective shell. Under a microscope, most limestones appear as pieces of various small or large organic structures. From tiny algae and foraminifera to bulky corals, most marine creatures use calcite for their hard parts. You may already sense the problem of separating limestones that are made of pieces of animal shells from the clastic rocks. As long as the rock is mostly calcite, it is limestone.

Quartz can precipitate in a variety of forms of microscopic SILICA. It may be called chert, flint, agate, jasper or chalcedony. All of these are quartz with a hardness of 7. CHERT is the term geologists prefer for the sedimentary quartz normally associated with limestone.

Rock salt, or HALITE, was deposited when ancient water bodies evaporated completely. GYPSUM also is deposited as a chemical precipitate after seawater undergoes extensive evaporation. Halite and gypsum often occur together and are the dominant "evaporite" minerals.

Chemical Sedimentary Rocks

gypsum	- $CaSO_4$ with H_2O
limestone	- mostly calcite ($CaCO_3$)
dolomite	- $CaMg(CO_3)_2$
salt	- halite (NaCl)
chert	- microscopic quartz (other names for microcrystalline quartz include agate, flint, jasper, chalcedony and opal)

IGNEOUS ROCKS: Igneous rocks are those that crystallize from molten rock (MAGMA). When magma begins to cool, either because it pours out on the surface or is chilled by surrounding rocks, the ingredients of the melt begin to form crystals and the melt freezes. The high temperature crystals form first, followed in a very precise order by the cooler minerals. The hot minerals often assume their distinctive crystal shapes, but the cooler minerals will have the shape of the spaces between the earlier crystals. Igneous rocks have interlocking crystals that are randomly oriented. For many reasons beyond the scope of this book, the high temperature minerals are segregated to form other associated rock types.

Igneous rocks are classified by two variables: mineral content and the size of the crystals. The size of the crystals is mostly determined by how fast the molten magma freezes. Deep-seated INTRUSIVE IGNEOUS ROCKS crystallize far beneath the surface of the earth and have large crystals that are easily seen with the unaided eye. EXTRUSIVE IGNEOUS ROCKS are those that pour out at the surface as LAVA and chill quickly, forming very fine crystals that are hard to identify with the unaided eye.

Remember that the rocks did not know about our classification, and some small bodies of intrusive rock that squeezed in between cool rocks may have small crystals. Other magma that poured out at the surface and made "lakes" hundreds of meters deep may cool very slowly and have visible crystals. If we know that the rock crystallized below the surface, we call it intrusive, and if we know if it formed at the surface, it is extrusive. Some igneous rocks that are removed from their source, say in a stream gravel, may be hard to identify correctly, even for the experts. For our purposes, if the rock doesn't fit nicely into its classification, it doesn't deserve a simple correct name!

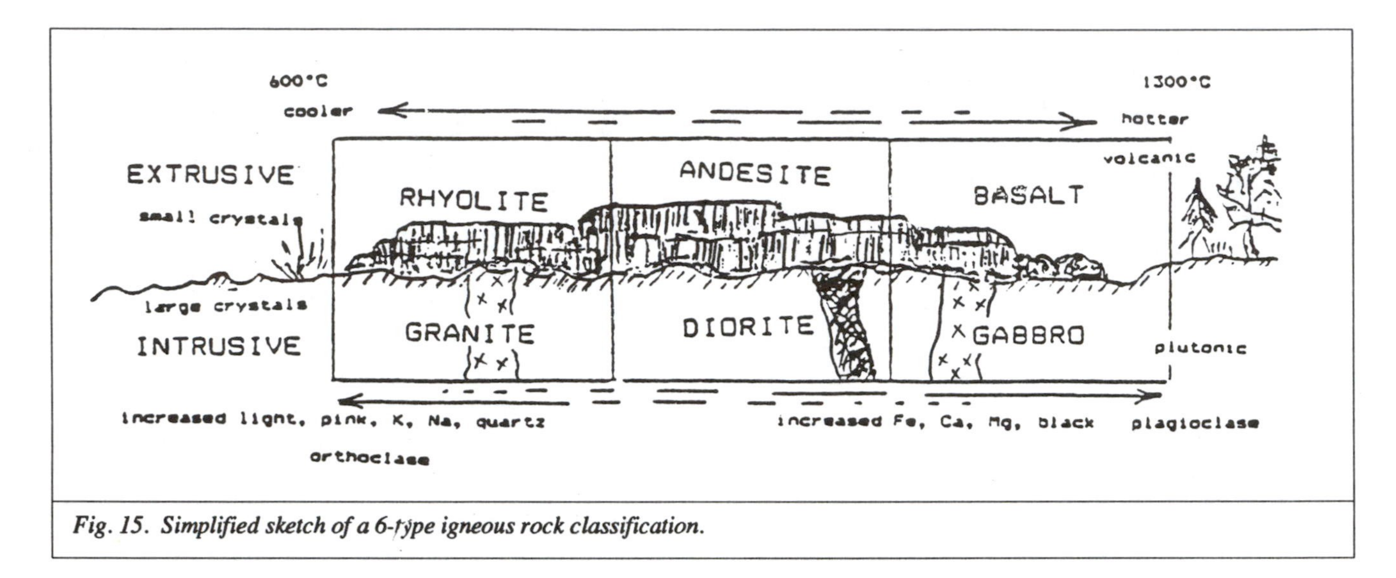

Fig. 15. Simplified sketch of a 6-type igneous rock classification.

There are igneous rock classifications with dozens of rock names. Ours will be simple, with only six basic rocks. If we draw a line that represents the surface of the earth, the rock names below the line are intrusives and the ones above the line are extrusives. High temperature minerals are in the rocks on the right side of the chart, and low temperature minerals are on the left. Minerals that crystallize at intermediate temperatures are in the middle of the chart. The intrusive rocks (bottom half of the chart) have coarse crystals that can be easily recognized with the unaided eye. Extrusive rocks on the top half of the chart crystallized quickly and have small or even microscopic crystals. The high temperature side (on the right) is enriched with dark or green, iron and/or magnesium-rich minerals that tend to be slightly heavier than the cooler minerals that are common on the left side of the chart. If plagioclase is present toward the right end of the chart, it will be mostly the Ca-rich variety. By "high" temperature, we mean about 1100° C. to about 1300° C. The cool side of the chart (left) includes minerals that tend to be light-colored or pink (orthoclase) and associated with quartz. If any plagioclase is present it will be the Na-rich plagioclase. Cool igneous rocks are those that crystallize near 600° C.

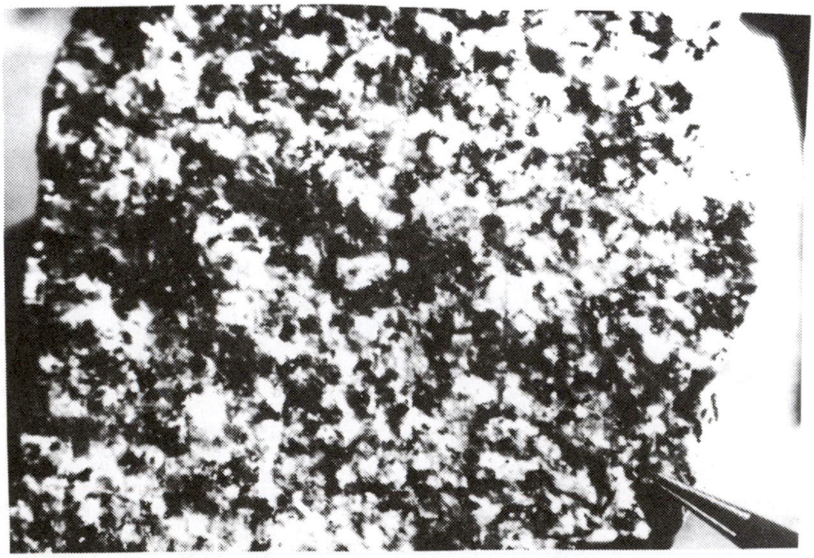

A few special characteristics of igneous rocks deserve to have names also. If a light-colored intrusive (granite) has spectacularly large crystals (over 3 cm. for example) it is called PEGMATITE. We should be able to see large crystals of quartz and orthoclase, and maybe a little mica and even a stray black mineral such

Fig. 16. This piece of granite has visible crystals that are randomly oriented and tightly interlocked. White crystals are quartz and orthoclase; black crystals are amphibole, biotite and possibly some magnetite (Fe_3O_4).

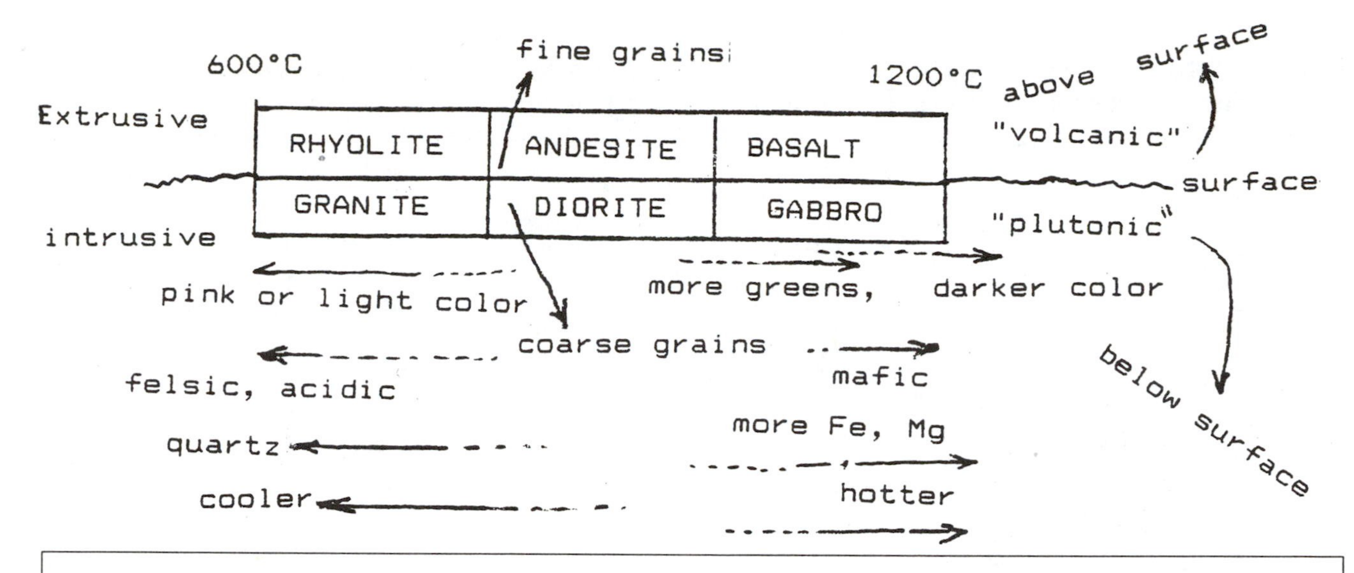

Fig. 17. The relationship of minerals, temperature, chemistry and grain sizes for the 6-type igneous rock classification.

as hornblende. The rock is "granite pegmatite." It is called granite because it is light-colored and intrusive. The word pegmatite recognizes the very large crystals. Gem topaz, tourmaline, ruby, sapphire, emerald, garnet and many others are found in pegmatites.

When lava cools with a lot of gas bubbles it may form a frothy type of rock that can actually float (PUMICE). The gas bubbles are technically called VESICLES, and if the rock has a lot of bubbles, but not enough to float, it is SCORIA. Volcanic glass is OBSIDIAN.

One last variety of igneous rock is the PORPHYRY. We know that this is a terrible word, but it defines a rock that has formed in two or more stages of cooling. The first stage was slow cooling that allowed a few of the hottest crystals to grow to significant size. Later, the mush of crystals and magma was chilled quickly, either by extrusion or perhaps the magma body vented at the surface and lost its volatiles or the pressure and/or temperature quickly changed. The liquid part of the magma freezes quickly, leaving a fine-grained matrix with a few large crystals. If the composition of the rock makes it an andesite, it would be an "andesite porphyry." If a coarse-grained rock has some even coarser grains, it may be "granite porphyry."

METAMORPHIC ROCKS: "Meta" (change) and "morphos" (form), or "METAMORPHIC ROCKS" are

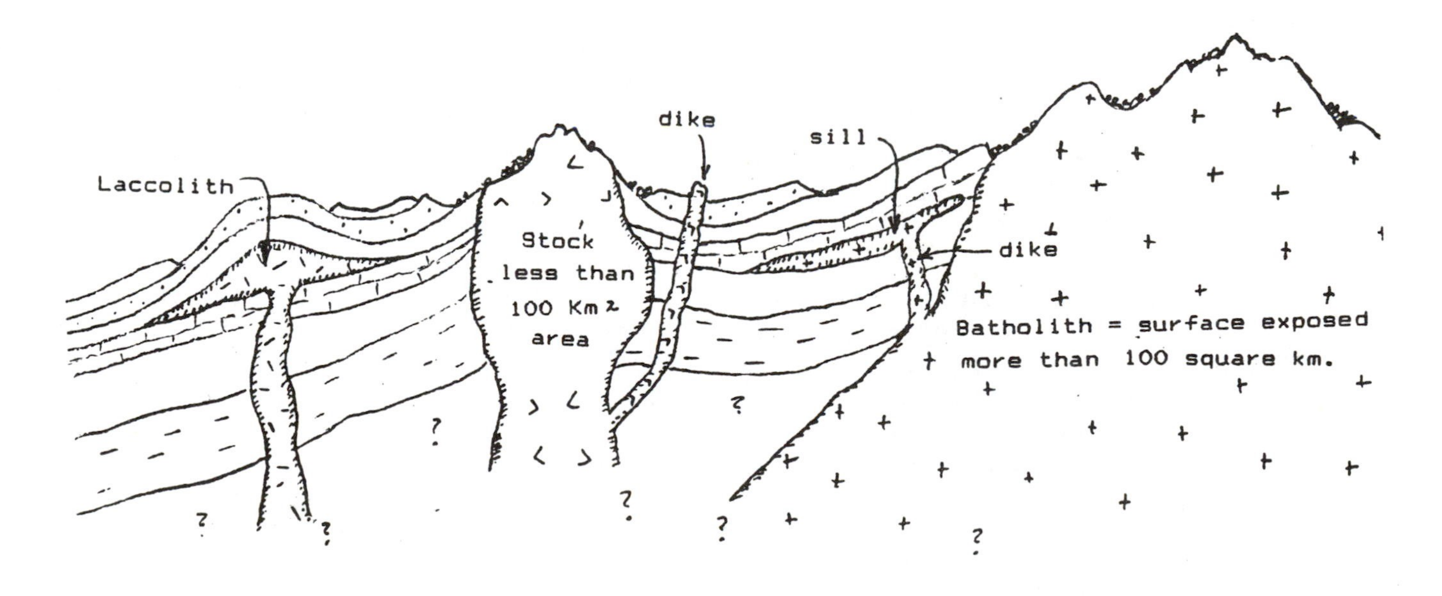

Fig. 18. Sketch showing the classification of intrusive igneous rock masses or "plutons".

those that have been changed in form by the addition of heat and/or pressure and or new chemicals. Oil wells have been drilled to over 10,000 meters in sedimentary rocks,

Fig. 19. A stream-rounded cobble of andesite prophyry. Though some of the grains are visible, it is extrusive lava. Three large phenocrysts of orthoclase probably grew to this size before the sample oozed out of a vent to chill quickly at the surface.

and the rocks are still basically sedimentâry rocks. Temperatures at those depths will be over 250 C but the rocks will be changed very little from rocks at 1,000 meters. To remelt rocks to make new igneous rocks would require burial to over 15,000 meters. Burial deeper than 10Km. but less than enough to melt the rocks can cause some minerals to recrystallize, but the rock might retain some of the sedimentary layering. Hot intrusions of magma, or special stresses from shifting plates of the earth's crust can also add enough heat and pressure to alter rocks. Hot water moving up along cracks in the crust may also heat or add enough chemicals to change shallow rocks.

GNEISS AND SCHIST are the common rocks of intense metamorphism. Gneiss will be composed of quartz, feldspars, micas and other minerals that are typical of sedimentary rocks, and the various minerals may become swirled bands or twisted patterns that usually are relics of the sedimentary layers. Things that caused lines in the sediments add different minerals that will recrystallize to make bands in the gneiss. Patterns in gneiss indicate that the rock behaved much like putty or partially stirred marble cake batter, but never completely melting and destroying the lines of the original sediments. The clay minerals will be destroyed in metamorphism. They are converted to the micas and a variety of other crystals, depending on the stuff contained in the original rocks and the chemistry of any invading fluid.

If there are enough clay minerals in the sediments, the resulting metamorphic rock will be dominated by mica flakes. When this happens, the rock is called SCHIST and it will split easily, parallel to the orientation of the mica flakes. In a metamorphic environment the micas may be parallel to the old sedimentary layers, or as they crystallize from the original clay minerals, they align themselves to the new stresses of the environment. It is common for relic sedimentary layers to exist in metamorphic rocks, but the cleavage (tendency to split along the micas) is at some angle to the old bedding lines.

If clean sandstones (mostly quartz) are metamorphosed the quartz grains often become the centers of new quartz crystals that enlarge to fill the porosity of the rock. This mosaic of interlocking quartz crystals is called QUARTZITE. Shadows of the former sand grains may still be visible in the rocks. Conglomerates of mostly quartz-rich rocks can be metamorphosed to quartzite that has the outline of the old pebbles clearly visible. Sometimes the pebbles are stretched almost to obscurity by the twisting of the rock while metamorphism was occurring. Again, the plastic behavior of near-melt conditions is evident.

When shale is metamorphosed to schist, it goes through two intermediate stages. When the clay minerals have been baked to a dense, dry, hard rock it is called SLATE before the mica grains are evident. The rock has good cleavage caused by microscopic mica, but the crystals are not apparent. When the mica crystals in slate become large enough for the rock to glisten, even before distinct crystals are seen, the rock is called PHYLLITE. When distinct crystals are seen and they control the cleavage, it is SCHIST. Other flat or rod-shaped minerals can also control the cleavage, as in an amphibole schist. If garnets are obvious in a mica schist it becomes a garnet-mica schist. The garnets are not part of the cleavage, but a distinct mineral helps refine the name.

MARBLE is what limestone becomes under metamorphism. It is still made of calcite, but the texture is recrystallized into a mosaic of sugary or larger crystals. Usually the color is bleached to white or pastel colors and if any impurities remain they will be in vague, swirled lines or patterns that give little hint of the former sedimentary layering. Some of the ornamental "marble" sold for

Table 1. Vocabulary for igneous rocks.

Term	Definition
ACIDIC	light colored, siliceous igneous rock
ANDESITE	extrusive rock of intermediate composition
APHANITIC	fine grained igneous rock, usually extrusive
BASALT	dark, mafic, high temp. volcanic rock
BASIC	high temp., low silica igneous rock
BATHOLITH	large intrusive, over 100 sq. km. exposed
CINDER CONE	small cone made of loose cinders, clinkers
COLUMNAR JOINTS	polygonal shrinkage columns in extrusive rocks
COMPOSITE VOLCANO	(see stratovolcano)
DIKE	discordant, tabular intrusive body
DIORITE	intrusive rock of intermediate composition
EXTRUSIVE	igneous rock that erupts at the surface
FELSIC	light colored igneous rock (opposite=mafic)
FIERY CLOUD	(see nuee ardente)
FISSURE FLOWS	widespread basaltic flows from long fissures
GABBRO	coarse grained, black intrusive rock
GLOWING AVALANCHE	(see nuee ardente)
GRANITE	light colored, siliceous intrusive, has quartz
IGNEOUS	derived from fire, molten source
IGNIMBRITE	rock formed from nuee artente
LACCOLITH	mushroom-shaped, small intrusive
LAHAR	mudflow of mostly volcanic debris
LAVA	extruded magma, either molten or frozen
MAFIC	high-temp. rock, low in silica (opposite=felsite)
NECK	dense rock that once filled the conduit of volcano
NUEE ARDENTE	hot, incandescent blast from stratovolcano
OBSIDIAN	volcanic glass, usually of felsic composition
PEGMATITE	very coarse-grained intrusive, grains over 2 cm.
PHANERITIC	coarse-grained igneous rock, usually intrusive
PHENOCRYST	an individual, large crystal in igneous rock
PLUG	(see neck)
PLUTONIC	coarse-grained, intrusive igneous rock
PORPHYRY	igneous rock with two sizes of grains
PUMICE	frothy volcanic rock; enough bubbles to float
RHYOLITE	extrustive felsic rock (granite equivalent)
SCORIA	dark lava with many vesicles, not enough to float
SHIELD VOLCANO	large, basaltic, low-angle volcano
STOCK	intrusive body smaller than 100 sq. km.
STRATOVOLCANO	steep, felsic cone of layers of lava and ash
TUFF	deposit of volcanic ash and dust
VESICLE	bubble hole in igneous rock
VOLCANIC BRECCIA	rock made of angular fragments of volcanic rock
VOLCANIC NECK	(see neck)
VOLCANIC PLUG	(see neck)
VOLCANIC ROCK	extrusive igneous rock; fine grained, aphanitic

decorating buildings is only well-cemented limestone. If the fossils and other grains are still distinct, the rock is not marble. However, if you are buying a polished rock to line the corridors of a great building, you just feel that you must use marble because it sounds more elegant. The Salem Limestone, of Mississippian age, and quarried from Indiana, is a popular stone that takes a beautiful polish. When sold as ornamental stone, it gets a better price as elegant marble than as cheap limestone. The hardness of marble is only 3, so if the texture is good, it is a preferred stone for

Fig. 20. This Precambrian gneiss was probably mudstones, dirty sand and/or siltstones before being buried several miles and cooked just short of melting. Relic sedimentary layers are the bands, distorted by twisting while very hot. The white is mostly feldspar and quartz, the black is a variety of minerals including amphibole and a little biotite. There is not enough mica to control the clevage.

Fig. 22. This marble tombstone at Telluride, Co. weathered less than 100 years to the point that fine engraving is not legible. With H=3, marble is not a hard rock.

Fig. 21. Shale (a) is soft, crumbly and splits in thin pieces due to the orientation of flat clay crystals. In slate (b) the clay crystals have recombined to make microscopic mica crystals that are oriented parallel to the table top, giving good clevage. The original bedding layers are still seen but are not parallel to the new clevage. Slate is hard. With more heat and/or pressure, the mica flakes grow and become visible, like fish scales, and give the rock the gleaming luster of schist (c). It has excellent cleavage, parallel to the mica flakes.

sculptors. Gravestones made of marble quickly lose their polish and deteriorate, especially in humid climates, because calcite is not a durable mineral.

This chart indicates where the **metamorphic** rocks come from. Sedimentary rocks (and igneous) of the left column are altered by heat and/or pressure and/or reactive solutions to produce the metamorphic rocks on the right column.

Previous rock:	Metamorphoses to:
limestone	marble
dolomite	marble
conglomerate	meta conglomerate
standstone (clean, mostly quartz)	quartzite
siltstone (mostly quartz)	quartzite
dirty sandstone (with feldspars, etc)	gneiss
dirty siltstone	gneiss
shale	slate, phyllite, schist
igneous rocks	gneiss, schist
coal	anthracite coal

Formations, Members, and Groups

When layers of sand, or some other distinctive sediment, are hardened to form rock, it becomes a distinctive sequence of rock layers, and gets the name FORMATION. The Dakota Sandstone is recognized all over Colorado and is one of the earliest sedimentary units of the Cretaceous Period in the state. It is the remains of a beach and the associated tideland sediments when a seaway moved in and covered much of the Western U.S.

Above the Dakota Sandstone is a great thickness of soft, gray shales which were deposited from the shallow sea. The gray shale contains many marine fossils, and is easily identified in most counties of the state. Such easily mapped sequences of distinctive rocks are called FOR-

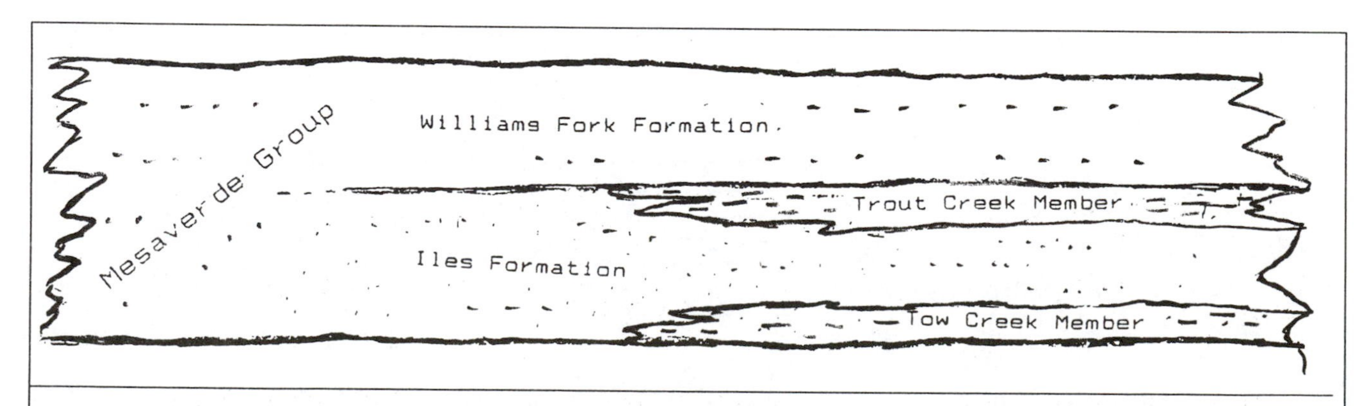

Fig. 23. This sketch shows the change in the Mesaverde units from west to east across northern Colorado. The group is neatly subdivided into formations, which develop members farther to the east.

MATIONS. The Dakota was named way back in 1862 and the TYPE SECTION, or first officially described outcrop, is a site along the Missouri River in Nebraska. The gray shales above the Dakota are called the Pierre (pronounced "peer") east of the Front Range, and Mancos in western parts of the state. Pierre was named for a type section at Pierre, SD. Geologists in western Colorado found gray shales similar to the Pierre, but they were not sure it was the same sequence because there was a mountain range separating the two areas of good outcrops. The sequence along the Mancos River, near the town of Mancos (by Cortez, in the southwest corner of the state) was formally named the Mancos Shale in 1899. We know now that the two formations were deposited in the same seaway, but the names have been recognized separately for so long, that both names are still proper. Formations are distinctive, mappable sequences of rock. They may be conglomerates, as with the Ohio Creek Conglomerate, which is a thin unit locally recognized in the upper Cretaceous section, or it may be a sequence of mixed rock types such as the Morrison Formation. The Morrison is a distinctive sequence, even though no single part of the sequence is worthy of an individual name. When parts of a formation deserve special recognition they are called formal MEMBERS of the formation. Most of the uranium found in the Morrison is found in the sandy lower part which is called the Salt Wash Member. Many formations are subdivided into members for local details. When two or more formations are grouped together for some useful reason, they may be given a formal GROUP name, as in the case of the Mesaverde Group, named at Mesaverde National Park, Colorado. Even igneous rocks and metamorphic rock units are given official formation names, such as the Pikes Peak Granite and the Idaho Springs Formation (schist and gneiss).

Unconformities and a Few Other Terms

In a normal sequence of sediments, Pennsylvanian-age material should be on top of Mississippian sediments. But most places have a very incomplete sequence of rocks. Deep oceans will have continuous sedimentation as long as the ocean is present. However, plates of the Earth's crust tend to wander about, tilting and sliding over one another from time to time throughout geologic time. When rocks are uplifted above the sea, instead of accumulating sediments, the existing rocks are stripped off by erosion. Pages of the geologic story are removed. We call such a break in sequence of rocks an UNCONFORMITY. There are three types of unconformity. In the case of the DISCONFORMITY, the sequences above and below the break are parallel, with no folding or other disturbance between the layers. There may be channels of erosion between the two sequences, but the bedding will be parallel. In the case of an ANGULAR UNCONFORMITY, the overlying layers are not parallel to the basal layers. This happens when a sequence of rocks is deposited, followed by twisting or uplift, even some erosion can occur, before the new units are deposited on top. A NONCONFORMITY is the sort of unconformity that has igneous or metamorphic rock on the bottom, covered with a sequence of sedimentary rock. Often geologists refer to

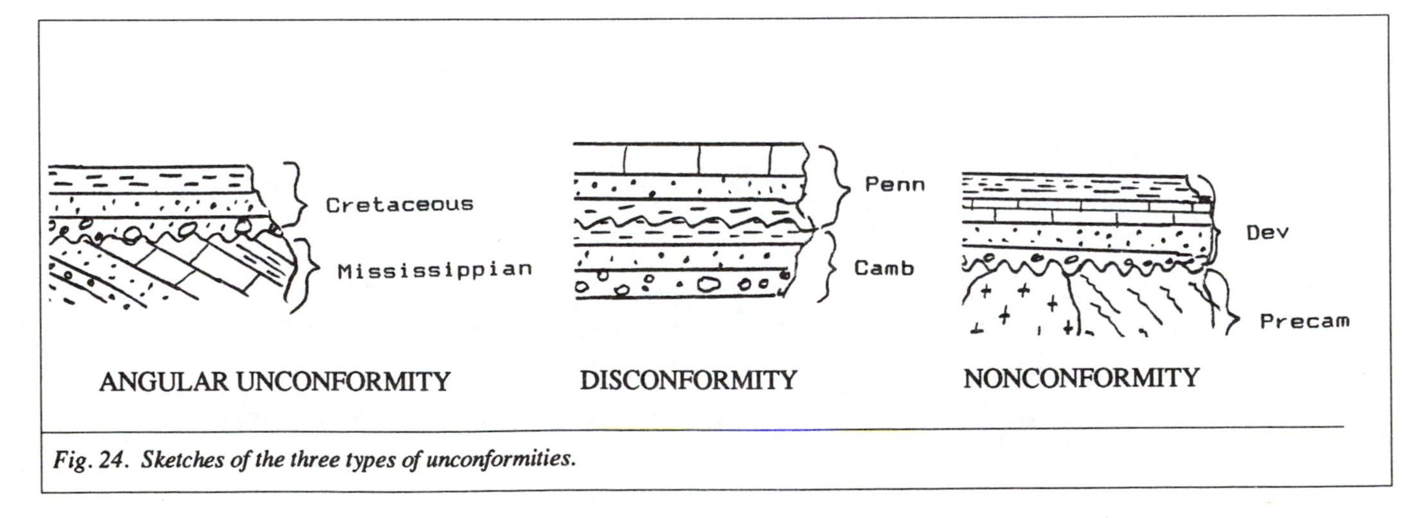

Fig. 24. Sketches of the three types of unconformities.

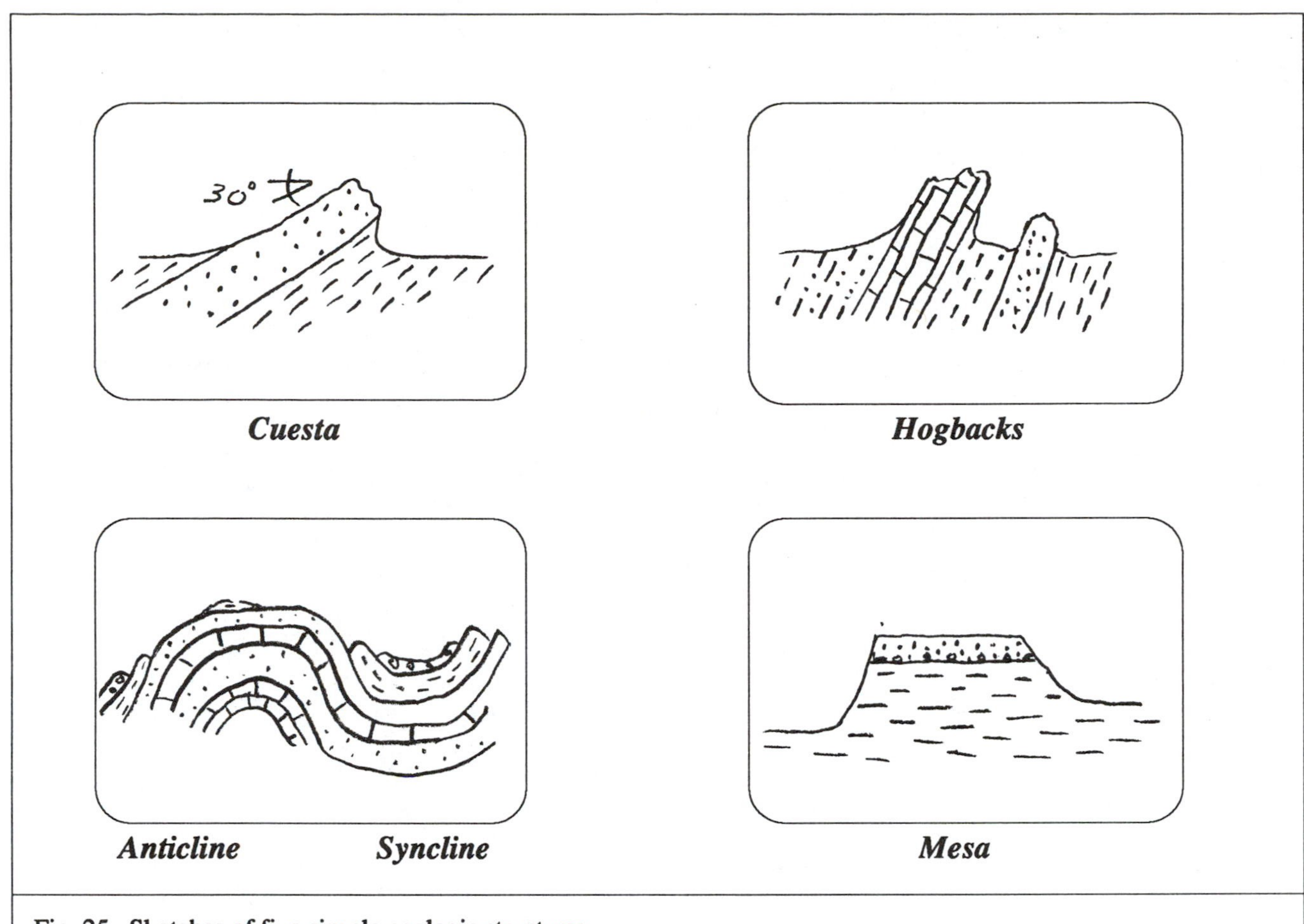

Fig. 25. Sketches of five simple geologic structures.

a "great unconformity" in most sequences in the Western States where sedimentary formations cover the ancient Precambrian rocks. The older rocks are usually metamorphic or igneous. Note in the diagrams (Fig. 24) that limestones are a series of brick symbols, sandstones are usually displayed as dots and shale is a pattern of dashes. Conglomerates are usually drawn as pebbles with sand dots. The symbol for an unconformity is a simple wavy line.

If a resistant rock unit is sandwiched between softer units, it can form a protective cap, making a MESA. When the sequence of soft and hard units is folded, the resistant unit stands out as a HOGBACK if it is tilted more than 30°, and a CUESTA if less than about 30°. The folds are called ANTICLINES if they arch upward, and SYNCLINES if they warp downward.

The Front Range

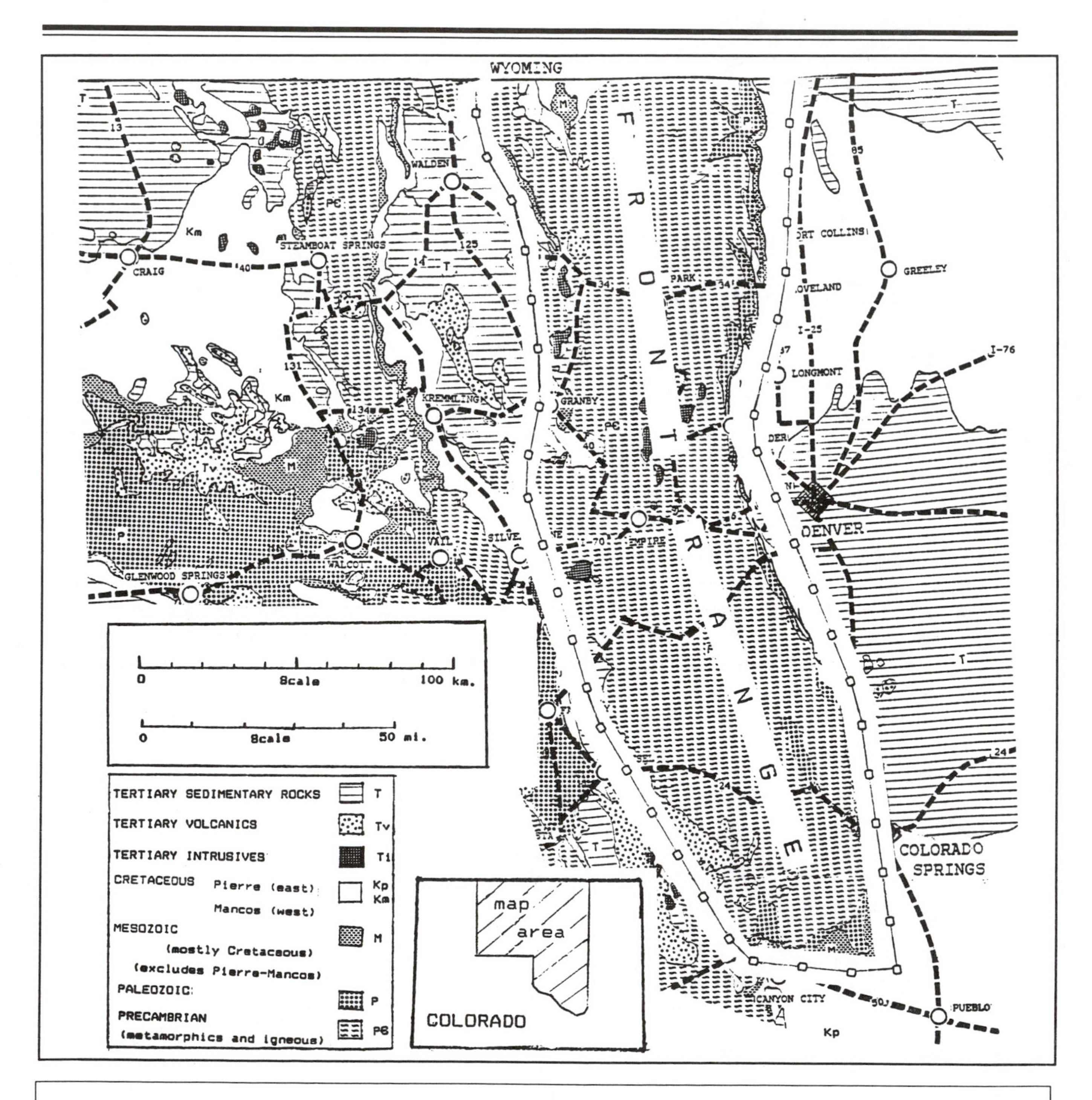

Fig. 26. Geologic map of north central Colorado, showing the Front Range.

Let's guess that a large number of people will become interested in the geology of Colorado when they arrive in the Denver area. If you are near Denver and the weather provides a view to the west, that impressive string of mountains is the Front Range. If you want to start some other place in the state, check the table of contents, remembering that we will gradually build our vocabulary of geology in the sequence of these book chapters. After starting with the Front Range, we will do the plains, then work our way westward, ending up at Dinosaur National Monument at the northwest corner of the state. Glaciers will be explained where they are important, and volcanoes will be explained in the area where they are significant.

Figure 26 is a simplified geologic map of the Front Range area of Colorado. The area outlined in Fig. 26 is about 12% of the total area of Colorado. Most of what is exposed at the surface is Precambrian material, about evenly divided between intrusive igneous and metamorphic rocks. The area east of the Front Range is the Great Plains, where the land is essentially flat at an elevation of about 1500 m. above sea level. (Mile High Stadium in Denver is exactly a mile above sea level, or 1600 m.). Longs Peak, west of Longmont, and Pikes Peak, west of Colorado Springs are "Fourteeners," meaning that they are over 14,000 feet above sea level (4200 m.). Both of these impressive peaks are composed of granite.

The range began pushing upward about 65 million years ago when the Laramide Orogeny ended the Mesozoic Era and began forming the entire chain of mountains that runs from Alaska to the southern tip of South America. In the case of the Front Range, the uplift was a little too rapid, resulting in the top of the range sliding off to the west (see Fig. 27). Near the town of Kremmling there are two small mountains with the base composed of soft Cretaceous shales of the Pierre-Mancos; and the top is hard, Precambrian metamorphic rock that rests atop the Cretaceous-about 50 km. from where it began its long glide off the summit of the early Front Range (Fig. 28).

On the east, or plains side of the Front Range, the hard, Precambrian block flopped outward, partly covering the softer materials underlying the plains. Fig. 29 is a simplified crossection of the Front Range at the approximate location of the I-70 route from Denver to Eisenhower Tunnel. The

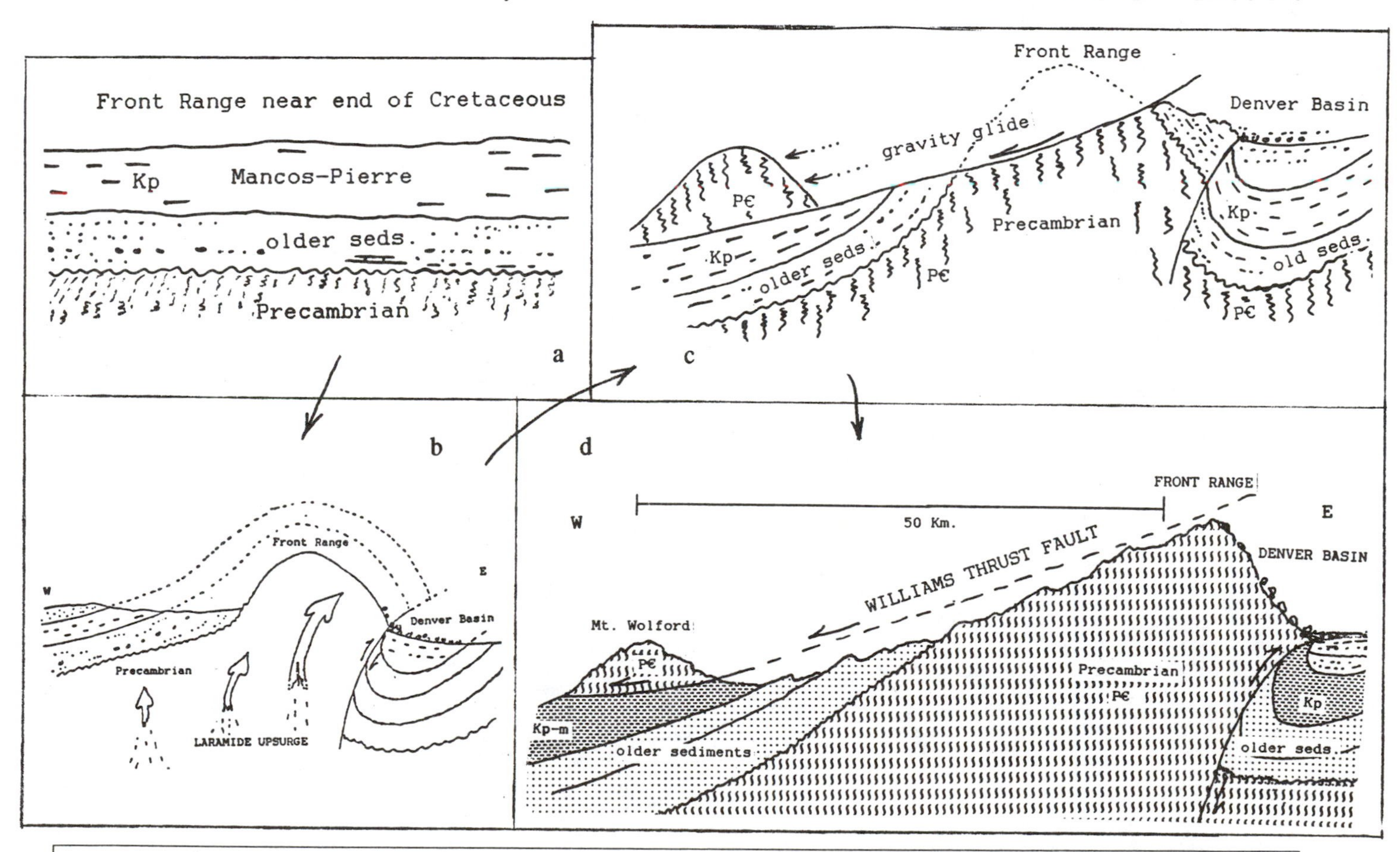

Fig. 27. Gravity gliding removes the top of some of the Front Range. In (a) the Pierre-Mancos marine sediments covered all of Colorado near the end of the Cretaceous Period. In (b) the Laramide Orogeny squeezes the basement rocks rapidly upward. Because of the excessive steepness, some of the top of the Front Range glides westward (c) and comes to rest near Kremmling (d), forming Wolford Mountain and Junction Butte.

crossection shows the relationship of the mountain escarpment to the Denver Basin, which abuts the east flank of the mountains. The basin is much steeper against the uplift than on the east edge of the basin where it extends gently eastward about 300 km. into Nebraska and Kansas. Such a downwarp is a syncline, and this one is very asymmetrical, and it plunges northward into Wyoming.

If we follow the profile of the mountain front in Fig. 29 from the mountain crest near Loveland Pass, the route is in Precambrian Idaho Springs metamorphic rocks. The route descends through the Silver Plume Granite, which has been dated at 1.45 billion years old. The Silver Plume has eaten part of the Idaho Springs, therefore we know it is younger than the metamorphic rock because the metamorphics had to be there first to be melted by the intrusive rock. It is difficult to date metamorphic rocks, but igneous rocks are rather easily dated. The Silver Plume also consumed

Fig. 28. Junction Butte, near Kremmling, has a slab of old Precambrian metamorphic rock resting atop Cretaceous marine shales of the Pierre-Mancos.

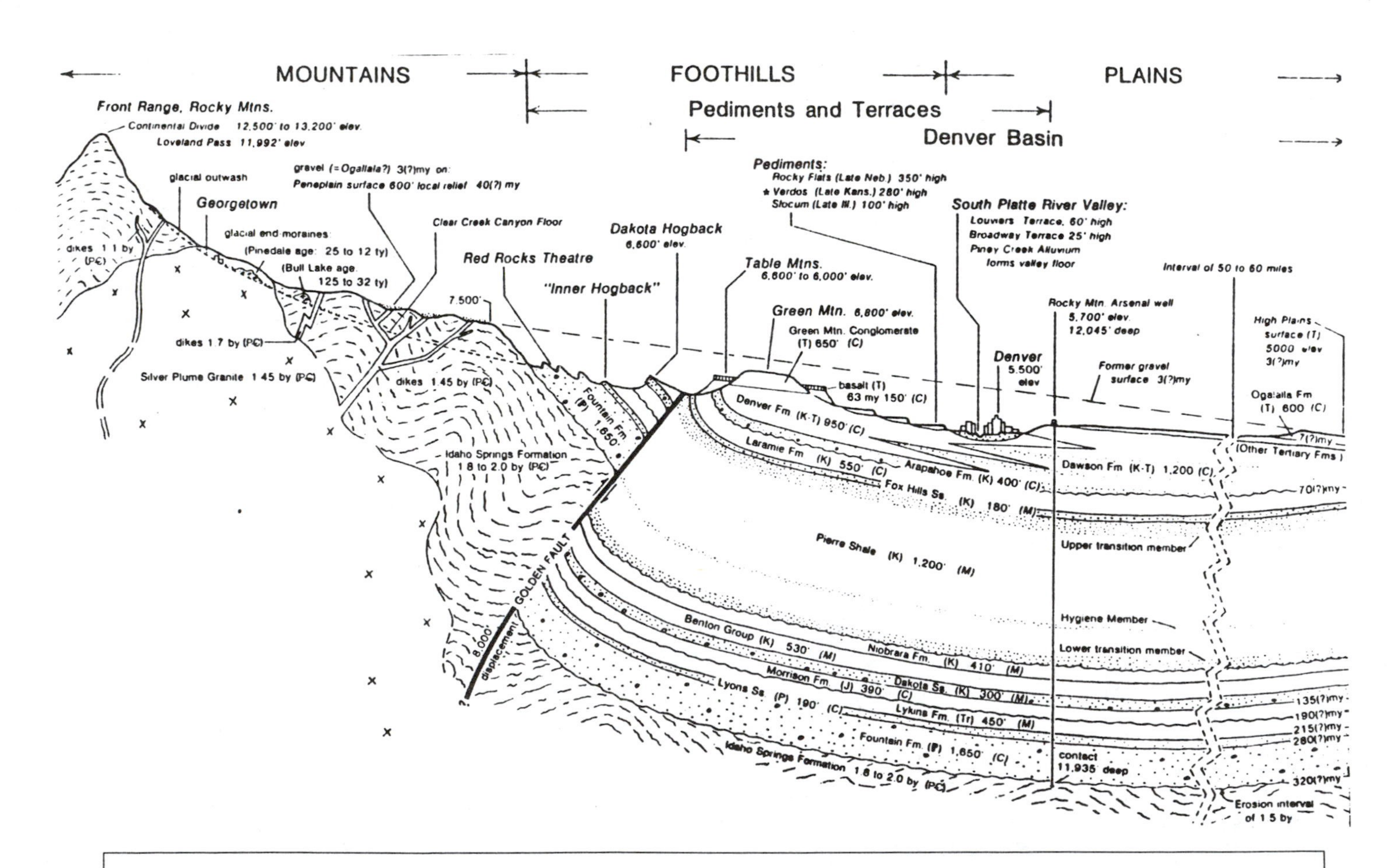

Fig. 29. A diagrammatic geologic cross-section of the Front Range, foothills and Denver basin along the route of I-70 (from H.E. Simpson, 1977, U.S.G.S., Denver).

dikes that were contained in the Idaho Springs, and the dikes have been dated at 1.7 billion years. We can extend the age of the metamorphics a little more-older than 1.7 billion years old. Note in the diagram that other dikes intrude the Silver Plume, and have been dated at 1.1 billion years old.

The Front Range is rather stable at present and the area is considered quiet, seismically. However, water disposal wells drilled near the Golden Fault (Fig. 29) stimulated a rash of small earthquakes in the 1960's that made Denver a very shaky place for a few years. When the water injections were halted, the earthquakes stopped. A renewal of water injections made an immediate response with new movement at depth, accompanied by more earthquakes. Injections have stopped; so have the quakes.

Fig. 30. Basalts dated at 63 million years old provide the cap for Table Mountains, overlooking the Colorado School of Mines at Golden, Colorado.

At Golden there are a couple of small mesas shown on Fig. 29. which are capped by basalts dated at 63 million years old. Most of the uplift of the Front Range occurred before the Table Mountain flows, because the basalt layers are not folded.

Rocks that were folded include the Dakota and Morrison formations that are exposed in a colorful road cut along Interstate 70, immediately west of Denver (Fig. 31). The Dakota includes the white sandstones that form the highest part of the rocks in Fig. 31. Toward the left side of the photograph the Morrison beds are the colorful, softer units. The U.S. Geological Survey has published a Professional Paper on this cut through the "Dakota hogback." The Denver Basin and plains are seen in the right background of Fig. 31. As described in the first chapter of this book, the Dakota is what remains from the beach deposits that were laid down when a shallow sea invaded the

Fig. 31. Interstate 70 cuts through the Dakota hogback west of Denver. A U.S. Geological Survey Professional Paper has been published on this exposure. Interpretive posters explain the scene on both sides of the highway.

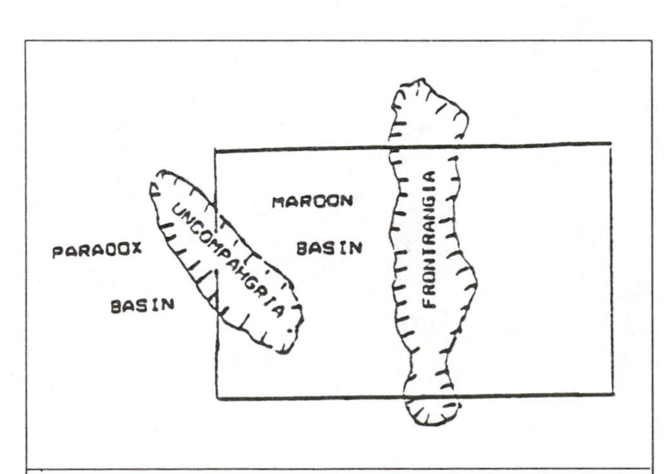

Fig. 32. The approximate location of two islands that bobbed up from the ocean that covered Colorado in Pennsylvanian time.

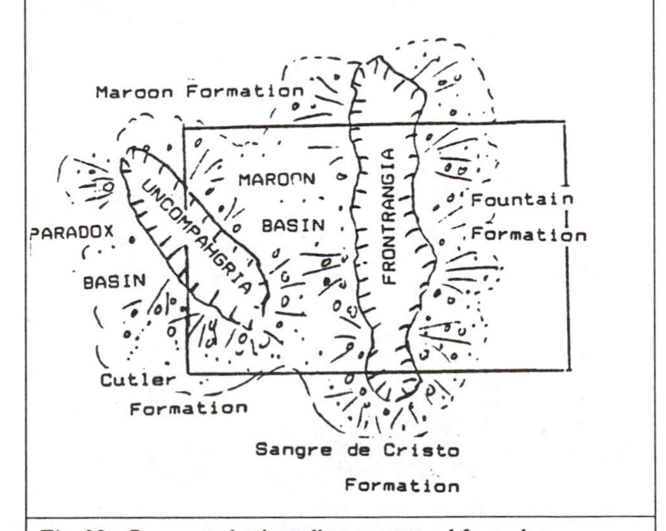

Fig. 33. Coarse, arkosic sediments spread from the two islands to become the Cutler (west), Maroon (center), Fountain (east), and Sangre de Cristo formations (south).

interior of the entire Western U.S. The sea ranged from Texas to Alaska. The Dakota is widespread and is usually overlain by the soft, olive-gray shales of the Pierre (Mancos in western Colorado).

Fig. 34. Flatirons of the Fountain Formation mark the foothills at Boulder, Colorado. NOAA's weather headquarters rest upon a Tertiary erosional terrace near the center of the photo.

During the Pennsylvanian Period an island we call Frontrangia arose in the area of the present Front Range. The island was a towering feature, not unlike our present Front Range, and the sediments which were stripped from all sides became the coarse, red arkoses of the Fountain Formation. The Laramide uplift pitched these red arkoses upward along the foothills of the Front Range in several spectacular exposures. Near Boulder the Fountain arkoses became the Boulder "flatirons" because they resemble a group of old flatirons used to press clothes on an ironing board (Fig. 34). The same red arkoses form the backdrop for Red Rocks Theater, located in the foothills just south of Interstate 70, between Denver and the town of Morrison. Morrison, of course, has the type section of the Morrison Formation, which is recognized in Colorado and four of its neighboring states.

Near Golden there are some nasty scars along the front of the mountains. People familiar with devastating earthquakes may think the scars are fault escarpments, but they are simply quarry sites for flagstones used in construction (Fig. 38). A distinctive Permian formation is the Lyons. This strongly layered sandstone splits evenly in layers abut 10 cm. (4 in.). These hard slabs have been used as construction "flagstones" and provide much of the distinctive beauty to the red buildings on the campus of the University of Colorado.

Fig. 35. Red, arkosic sandstone and conglomerate of the Fountain Formation (Pennsylvanian age) are the backdrop for Red Rocks Park west of Denver.

Fig. 36. At Red Rocks a bronze plaque identifies the nonconformable contact between the Precambrian Idaho Springs metamorphics and the overlying Fountain arkoses of Pennsylvanian age.

Fig. 37. Arkosic conglomerate at the base of the Fountain Formation includes a coarse cleavage fragment of orthoclase feldspar, marked by the pen.

Fig. 38. The foothills near Golden are scarred by stone quarries.

Fig. 39. Lyons Formation near the town of Lyons, Colorado.

Fig. 40. Table tops cut from the easily-split "flagstones" of the Lyons Formation.

Fig. 41. Pink flagstones from the Lyons Formation. Similar slabs were used for many buildings on the campus of the University of Colorado.

The Lyons, and units above and below, are responsible for an unusual city park, the Garden of the Gods. In 1991 the city of Colorado Springs managed this cluster of hard rock ribs and allowed tourists free access to the park without any admission charge. The park has remarkable spires of silica-cemented pink sandstone that stand up to 80 meters above a wooded park area. Soft layers between the hard ribs have been removed by erosion.

Whenever there is steep topography in limestones, there is always the chance for caverns to develop. Near Colorado Springs the Devonian and Mississippian limestones have been dissolved into a set of caves called the Cave of the Winds. These are not very spectacular compared with the more famous caves around the nation, but Colorado has very few good caverns, and none of which are as close to large populations as is the Cave of the Winds. The caverns are exploited a little, but they are interesting. At Manitou Springs is a good exposure of the Cambrian Sawatch sandstones resting nonconformably on the Precambrian.

Fig. 42. This balanced rock at Garden of the Gods is part of the free park area, administered by the city of Colorado Springs.

Fig. 43. A few ribs of pink, resistant sandstone were twisted to a vertical orientation below Pikes Peak at Garden of the Gods.

Fig. 44. Cave of the Winds has been eroded from Mississippian and Devonian limestones in the foothills west of Colorado Springs. The entrance is shown in the upper left.

Fig. 45. Cambrian Sawatch sandstones lie nonconformably atop the Precambrian metamorphics at the Manitou Springs exit off U.S. 24.

Pikes Peak

Pikes Peak dominates the south end of the Front Range. It is the high point of the Pikes Peak batholith, which is a circular mass of coarse, pinkish granite, over 60 km. in diameter, located just west of Colorado Springs. Tourists make a big fuss over Pikes Peak as it is an impressive fourteener, visible almost as far away as Kansas. There is a town called Firstview on the plains just 45 km. inside the east border of Colorado. Tourists entering the state on U.S. 40 get their first view of Pikes Peak from that town, which is approximately 220 km. distant (136 mi.). When the weather cooperates, the view is hard to believe when the distances are known. Usually the peak has some snow on the summit, and after travelling a few thousand kilometers, especially the preceding 650 km. across Kansas, it is spectacular. It is named for Zebulon Pike, a U.S. Army captain who tried to climb it about 1806. He thought it was about 5400 m. high (18,000 ft.), and when he failed in his attempt to climb it he said that it would never be climbed by man.

At 4233 meters, Pikes Peak is only number 31 among 54 fourteeners in the state, and California's Mt. Whitney is more than 115 meters higher. Alaska has peaks over 6,000 meters (20,000 ft.).

Nevertheless, tourists like Pikes Peak, so this text can explain what it is. The pink granite is rich in coarse orthoclase feldspar and much hematite staining in most areas. Age of the granite is Precambrian and the massif bobbed up during the Pennsylvanian and had all the overlying sedimentary rock stripped off by erosion. An apron of arkose was deposited around the base during the time of the island Frontrangia, but the mountain stayed subdued from late Permian until its fierce uplift of the Laramide, to close the Mesozoic Era. Erosion is tearing away at the batholith, but it is pretty hard rock, and will no doubt last longer than most of our readers.

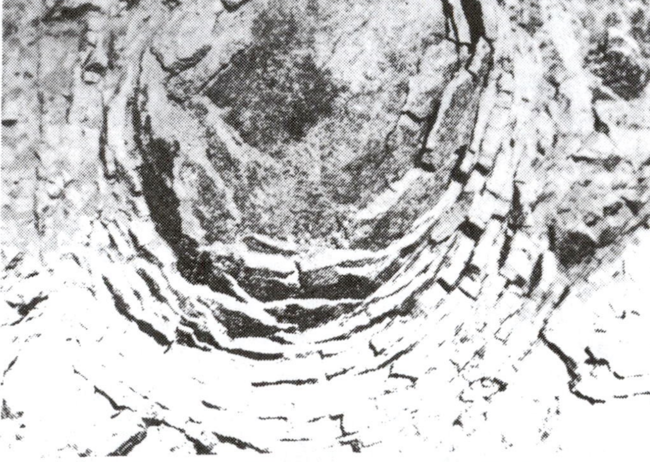

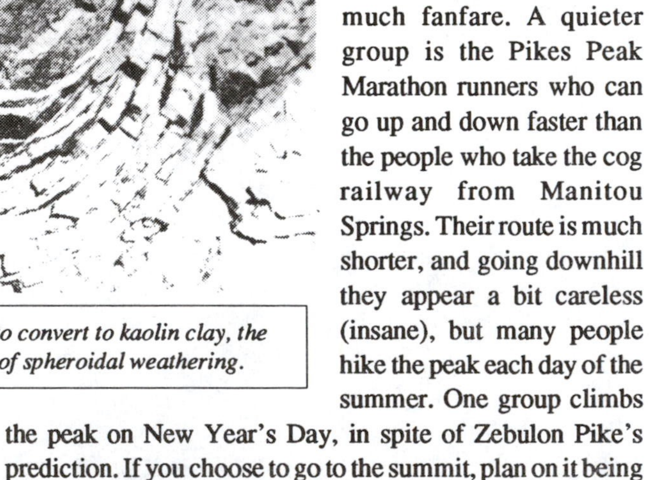

Fig. 46. When feldspars begin to convert to kaolin clay, the change in volume causes shells of spheroidal weathering.

Pikes Peak is a good place to introduce a distinctive weathering feature of igneous rocks called SPHEROIDAL WEATHERING. When igneous rocks begin to weather, one of the first things to happen is the feldspars begin to deteriorate to kaolin clay. When this happens, the volume of the feldspar first expands, slightly, to accommodate the addition of water to the clay structure. The expansion weakens an outside crust of the rock, allowing a shell to pry itself away from the core of the rock. Figure 46 shows the remains of a 0.5-meter boulder of igneous rock that started as a block, separated by joints from its neighboring blocks. Repeated stages of spheroidal weathering have reduced the block to a nest of shells. In Figures 47 and 48 the shells of weathering are able to fall away from the outcrop, leaving the remaining mass with the rounded surface typical of igneous rocks. In Fig. 49 Pikes Peak Granite is exposed in a road cut along Highway 24 near Divide. In this view, the top half of the photo shows weathered, coarse granite, and the underlying shell is still very hard, even though it is separated by lower shells. It is usually possible for a good geologist to spot igneous rock from a considerable distance, even a kilometer or more.

The summit of Pikes Peak is monotonous granite, but there are a variety of precious stones collected from and adjacent to the batholith. A good, unpaved road reaches the summit and it is one of the favorite routes for international road racers, who tackle it each year with much fanfare. A quieter group is the Pikes Peak Marathon runners who can go up and down faster than the people who take the cog railway from Manitou Springs. Their route is much shorter, and going downhill they appear a bit careless (insane), but many people hike the peak each day of the summer. One group climbs the peak on New Year's Day, in spite of Zebulon Pike's prediction. If you choose to go to the summit, plan on it being cold. A pleasant mild day in September, at Colorado Springs, may have a blizzard with gale-force winds on the top. The restaurant and concessions on the top are warm enough, but it is cool on the outside.

Fig. 47. Pikes Peak granite exhibits the rounded surface typical of weathered, intrusive rocks.

Fig. 48. Characteristic Pikes Peak granite is exposed near Cripple Creek, Colorado.

Fig. 49. Granite exposed in a road cut along U.S. 24. The top half is strongly weathered, the bottom is much less altered.

Fig. 50. The east shoulder of Pikes Peak as viewed from the cog railway that climbs the peak from Manitou Springs.

Fig. 51. Above timberline the Pikes Peak granite is shattered by frost action, modifying the spheroidal weathering habit of igneous rocks. Here the cog railway rounds the south side of the mountain.

Fig. 52. Hundreds of square kilometers of the Pikes Peak batholith can be seen from the cog railway.

Fig. 53. Pikes Peak may be the most famous "fourteener" in Colorado, but the state has 30 higher peaks.

Florissant Fossil Bed National Monument

Florissant Fossil Beds National Monument is one of the best kept secrets of Colorado. It is secreted in a remote area of the Pikes Peak batholith on a dirt road and buffered by Cripple Creek, which draws away most of the unenlightened tourists. The official visitor count in 1991 was only 37,000! It is a world class deposit of fossil insects and plants, including giant *Sequoia* trees, similar to those of the Sequoia National Park in California-only the Colorado variety cannot burn!

Marked nature trails lead from the visitor's museum to a dozen or so stumps of trees that are up to three meters in diameter, and must have approached a hundred meters in height.

In 1874 the first insects and leaves were recovered at the site by Dr. A.C. Peale of the U.S. Geological Survey. The variety was so remarkable and the preservation so superb that in 1969 the U.S. Department of Interior withdrew the area from collecting by the general public. As a national preserve, the area can be studied, and fossils can still be extracted, but only under carefully controlled permits. Researcher F. Martin Brown reports that by 1988 1,271 species of insects, 8 mollusks, 3 fish and 3 birds had been identified at Florissant, plus 114 species of plants.

Fig. 54. Petrified Sequoia stumps at Florissant Fossil Beds National Monument are wrapped in steel bands to delay the crumbling action of weathering.

Why is Florissant so special? If one recalls the junior high school science formula for making good fossils, the answer is clear. The recipe is (a) a quick burial (b) no oxidation or decay, and (c) preservable hard parts on the biota. At Florissant steps (a) and (b) are so complete that (c) can be ignored. Insect wings, leaves, larvae, and other delicate items are recovered intact.

Back in Oligocene time a verdant forest of giant *Sequoia* trees shaded a small, lava-dammed lake at the site. In those days it was

Fig. 55. Oligocene insects are among the fossils at Florissant.

1000 meters above sea level and the banks were marshy with a wide variety of plants and animals flourishing. Then a nearby volcano, probably to the west, belched an acrid cloud of smoke, cinders and ash. Either many volcanoes joined the chorus or a single vent erupted repeatedly. At any rate, the lake received a suffocating layer of ash. The fossils are not confined to a single layer, so it appears that the environment recovered a number of times, only to be smothered again and again.

The bases of the *Sequoia* trees were buried three to six meters deep in ash. The trees died with only the bases buried. In time the tops rotted away in a barren, desert-like desolation, similar to the north side of Mt. Saint Helens, a condition which continued more than a decade after the 1980 eruption. Even in rainy Washington State, soil is rebuilt very slowly, and in 10 years few large shrubs had reclaimed their domain.

Today some of the buried fish, insects and plant litter at Florissant are exquisite fossils. The *Sequoia* bases are now replaced by quartz which was leached from the ash and redeposited in the wood cells. The wood grain is faithfully preserved, and the loose pieces look temptingly like firewood. However, an axe of hardness 5.5 would be ruined trying to split quartz logs with a hardness of 7.

The park service has exhumed the stumps for better observation by the visitors. As soon as the restraining sediment was removed, the stumps began to deteriorate-not as logs would decay, but more slowly, like a ledge of sandstone. Now the stumps are fenced in to protect them from the souvenir seekers, and carefully wrapped in steel bands to delay their inevitable demise to the agents of erosion.

Cripple Creek

Cripple Creek can be seen from Pikes Peak, and is located about 15 km. southwest of the summit. In 1890 Cripple Creek was probably the greatest gold camp of the entire world. At the Cresson Mine, workers broke into a room lined with gold and silver minerals. Stories about that "Cresson Vug" are among the most unbelievable of mining history. About 400 tons of gold were taken from the district and the population grew to an estimated 20,000 to 30,000 people. The boom lasted until World War I when all the nation's gold mines were closed in order to release strong young men for the military. Some of the mines became active again, but were closed again in World War II for the same reason.

Fig. 56. The hills at Cripple Creek and Victor are pock-marked by old mines. Ore waste piles are being re-worked for new gold.

In 1992 there was little activity in the old underground workings. Most of the revenue for the Cripple Creek district was from tourism. Gambling became legal at Cripple Creek in 1991, so the economy of the area has taken a new direction. However, another quiet and very lucrative operation is underway at both Cripple Creek and the sister town of Victor. The old mine wastes are carefully piled and leached with a cyanide solution in a process called HEAP LEACHING. The old mine dumps are especially amenable to leaching because the rock has already been crushed and the ore and the gold are mostly the size of fine dust. After the cyanide solution has percolated through the heap, it is deflected by plastic sheets into a pond of "pregnant solution" which is then treated to extract gold and silver from the solution. In the heyday of the big mining camps, a lot of the fine gold was discarded because the big operations could make plenty of profit on the coarser gold, or with the higher grade ore deposits. But in those days, an ounce of gold was only worth $20 (70 cents a gram). When the U.S. dropped the gold standard around 1970, the price of gold jumped to nearly $1000 an ounce ($35 a gram). There has been a tremendous jump in the number of active gold mines in the U.S. since 1970. The price has stabilized below $400 an ounce ($14 a gram) for a decade, but the technology of the heap leaching has dominated gold mining in the U.S. in recent years.

Fig. 57. Cyanide leaching is being used on this tailings pile to extract some of the fine gold that was missed in the boom days at Cripple Creek

Glaciers

About two million years ago the earth began a very unusual rhythm of ice cap surges near the poles and down the higher peaks in the warmer latitudes of the globe. Those two million years comprise the Pleistocene Epoch, or ice age. The only two other episodes of significant world-wide glaciation occurred in the Precambrian and at the close of the Permian. Today's climate is a little cooler than the typical interglacial temperatures of the Pleistocene. Our climate will likely get a little warmer before it goes to another cold cycle. There is no guarantee that it will get colder or warmer. The glacial stages are named from the U.S. locations where each was first recognized as a separate glacial deposit (Fig. 58). Europe has its own stage names.

In Colorado any mountain above 2700 meters (9,000 ft.) probably had a permanent snow pack during the Pleistocene or "ice age." On the north, or shaded, side of the peaks above 2700 m., glaciers formed whenever the catchment area was large enough and the slope was steep enough for the ice to move. Higher mountains received more snow and glaciers could form on all sides. The tall peaks were intensely scoured by moving ice to make the Colorado Rockies fit the jagged model of glaciated peaks so familiar in the Alps or Himalayas and the higher American Rockies of Canada, Alaska and the Andes. A few snow fields remain. The U.S. Geological Survey topographic maps show approximately 50 ice fields and/or genuine glaciers.

Timberline in Colorado is approximately 3450 m. (11,500 ft.), and the lack of trees tends to exaggerate the starkness of the peaks. If our geologic calendar had not ended with the ice age, most of the world's famous peaks would not be nearly as exciting as they are.

In order to appreciate the impact of glaciers in Colorado, it is important that we recognize the features formed by the ice. The following discussion of glacial scour features and glacial deposits should prepare the beginner to be an astute observer.

When a snowpack is compressed to ice and begins to slide

down a mountain to become a valley glacier, the moving part pulls away from the snow cap that remains on the peak. The crack, or crevasse, that forms at the upper limit of the glacier is given the special name of BERGSCHRUND. As even more snow falls, the bergschrund may fill with snow or debris-laden avalanches, but the constant movement of the glacier continually re-opens the bergschrund. The snow cap also receives more snow which makes it stretch and cave-off avalanches into the bergschrund. Rocks that fall with the avalanches, into the bergschrund, become the scouring material that stays imbedded in the solid ice, making the ice a gigantic tongue of sandpaper, with the sand grains possibly as large as a house.

The scoured basin at the head of the GLACIAL TROUGH (ice-scoured valley) is a CIRQUE. If a lake fills the basin it is called a TARN. A HORN is a sharp mountain peak with cirques on three or more sides. The classic horn is the triangular-shaped Matterhorn of the Swiss-Italian border. A knife-like ridge that separates two glacial troughs is called an ARETE. Rock debris embedded in a moving glacier can polish or scratch the underlying rock surface, causing GLACIAL POLISH or STRIATIONS (scratches). Rock material deposited at the end of the glacier becomes a TERMINAL MORAINE. If debris is piled at the flank of a glacier, it becomes a LATERAL MORAINE when the ice melts. Individual rocks in the moraine (or dropped by the ice in any foreign location) are called ERRATICS. Rocks embedded in the glacier may develop flat, scratched surfaces as they scrape along the side or bottom of the trough. A scratched boulder found in a glacial deposit is called a FACETED BOULDER. If a rock, imbedded in the ice, scratches along and then hits an obstruction, it may twist in the ice and present a new surface for scratching, and there will be more than one plane of faceting on the rock. Material left by a retreating glacier is GROUND MORAINE. If a retreating glacier pauses in its retreat and holds the same position (forward motion balanced by melting) a RECESSIONAL MORAINE may form.

When the ice age began, ice sheets (CONTINENTAL GLACIERS) pushed southward from heavy snow centers in Canada while the VALLEY GLACIERS expanded down the mountains of the more temperate latitudes. The warm cycles, between glacial advances, were mostly warmer than our modern climate. It was enough warmer that alligators lived in the Great Lakes area of the United States during interglacial episodes.

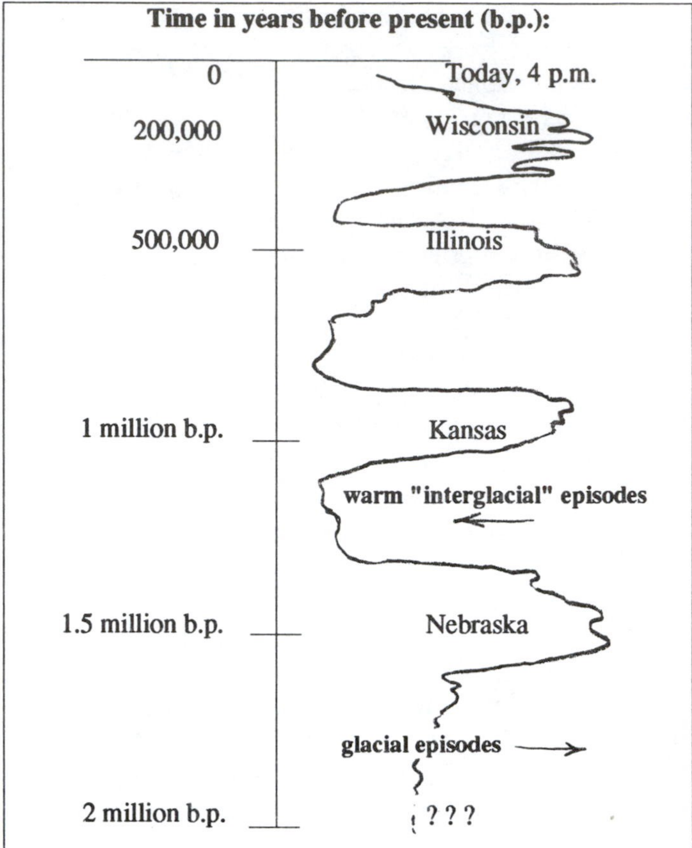

Fig. 58. A chart showing the swings to cold weather (right) and warm interglacial episodes (left) during the ice age.

As glaciologists deciphered the position of each advance, they gave names to the advance based on the site where the advance was first identified as a separate surge. For example, the Nebraskan Stage was the earliest of the Pleistocene advances, based on exposed debris found in Nebraska. The last one was first identified in Wisconsin, so it is called the Wisconsin. It is the easiest to decipher, and four shorter surges are recognized in the Wisconsin Stage.

Two common glacial features are the DRUMLIN and the ROCHE MOUTONNEE. The drumlin is a streamlined hill of till, but the roche moutonnee is a scoured knob of bedrock. Streamlining of the two is reversed, as shown in Fig. 61. Also, the lower end of the glacier is called the SNOUT, and the length of the glacier is determined by the balance between new snowfall and melting. The upper end, where

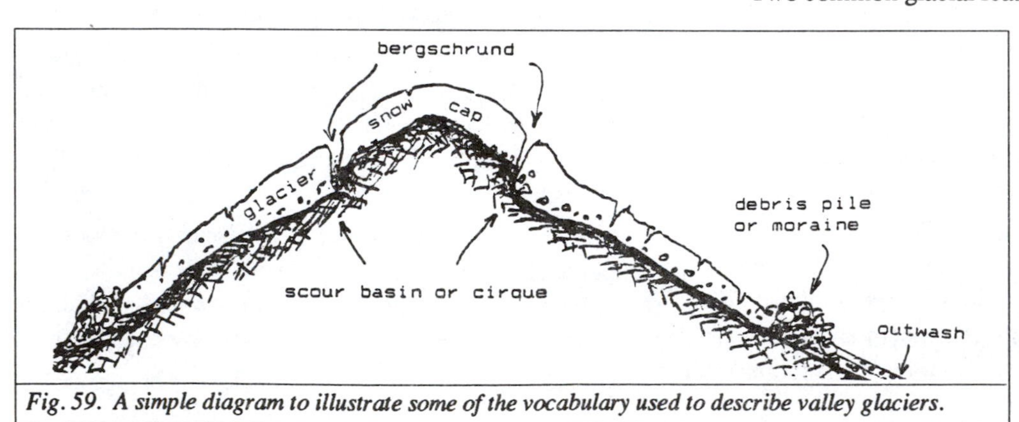

Fig. 59. A simple diagram to illustrate some of the vocabulary used to describe valley glaciers.

snowfall exceeds melting is called the ZONE OF ACCUMULATION and the lower end, where melting exceeds snowfall is called the ZONE OF ABLATION. As the ice carries debris to the moraine, meltwater is constantly shed by the system. This profusion of water flows around, over and through the moraine and distributes much of the glacial debris as an apron of sediment downstream from the glacier called an OUTWASH PLAIN. Outwash may be destroyed by post-glacial erosion as the stream continues to erode its valley. Often a remnant OUTWASH TERRACE may look like a moraine. Old moraines may look like an isolated piece of an outwash terrace. Usually one can tell the difference between an outwash terrace and a moraine, as the outwash is stream-deposited and will be stratified, more or less sorted in sizes, and most of the stones will be partly rounded. The characteristics of glacial material are three "un's:" unsorted, unrounded and unstratified. Any material deposited directly by the ice of a glacier is called TILL. The table below summarizes the vocabulary of glaciers.

Fig. 60. A valley glacier slid into this area from the lower left corner of the photo. Debris was dumped at the end, or snout of the glacier forming a morainal loop, with lateral moraines on the sides and a terminal moraine around the end. In the background the outwash material was spread, apron-like, in front of the moraines by streams that breached the moraine dam. This is Wallowa Lake, Oregon, but looks similar to many moraine-dammed lakes in Colorado.

Fig. 61. A sketch to show the difference between a drumlin (deposit) and a roche moutonnee (scoured bedrock).

arete	knife-like ridge between two glacial troughs
bergschrund	crevasse at the head of glacier (pull-away portion)
cirque	spoon-shaped basin at head of glacial trough
crevasse	deep crack in a glacier
drift	all glacial deposits, from water or ice (see till)
drumlin	streamlined hill of till (forms under glacier)
erratic	a stray rock, dropped by glacier
esker	snake-like ridge deposited by stream under glacier
firn	granular (remelted-recrystallized) snow
firn line	color change on glacier (new snow on old, dark snow)
glacier	dense ice that lasts through summer and moves
ground moraine	scattered debris from retreating glacier
hanging valley	tributary valley with floor higher than main valley
horn	jagged peak with cirques on 3 sides
kettle	pond formed by melting ice block in till or outwash
lateral moraine	moraine left at the side of a glacier
loess	wind-laid soil, common after outwash is blown around
moraine	ridge of till deposited by edge of a glacier
mutton rock	roche moutonnee
outwash	flood plain of streams flowing from snout of glacier
recessional moraine	moraine ridge formed during retreat of glacier
roche moutonnee	ice-scoured mound in bedrock
rock glacier	tongue-shaped pile or lobe of moving talus
striations	scratches
tarn	lake in a cirque
till	unsorted, unstratified, unrounded glacial deposits
terminal moraine	ridge of till at farthest advance of a glacier
U-shaped trough	glacially scoured valley
whaleback	roche moutonnee (surrounded by water)
zone of ablation	portion of glacier with more melt than snowfall
zone of accumulation	portion of glacier with more snowfall than melt

Rocky Mountain National Park

Rocky Mountain National Park gets more tourist visits than all the other Colorado national parks and monuments combined. What did the 2.8 million visitors in 1991 see? Glacial features! The park was established in 1915 and drapes over the Continental Divide between Estes Park and Lake Granby. The Mummy Range, Neversummer Range, Longs Peak and Indian

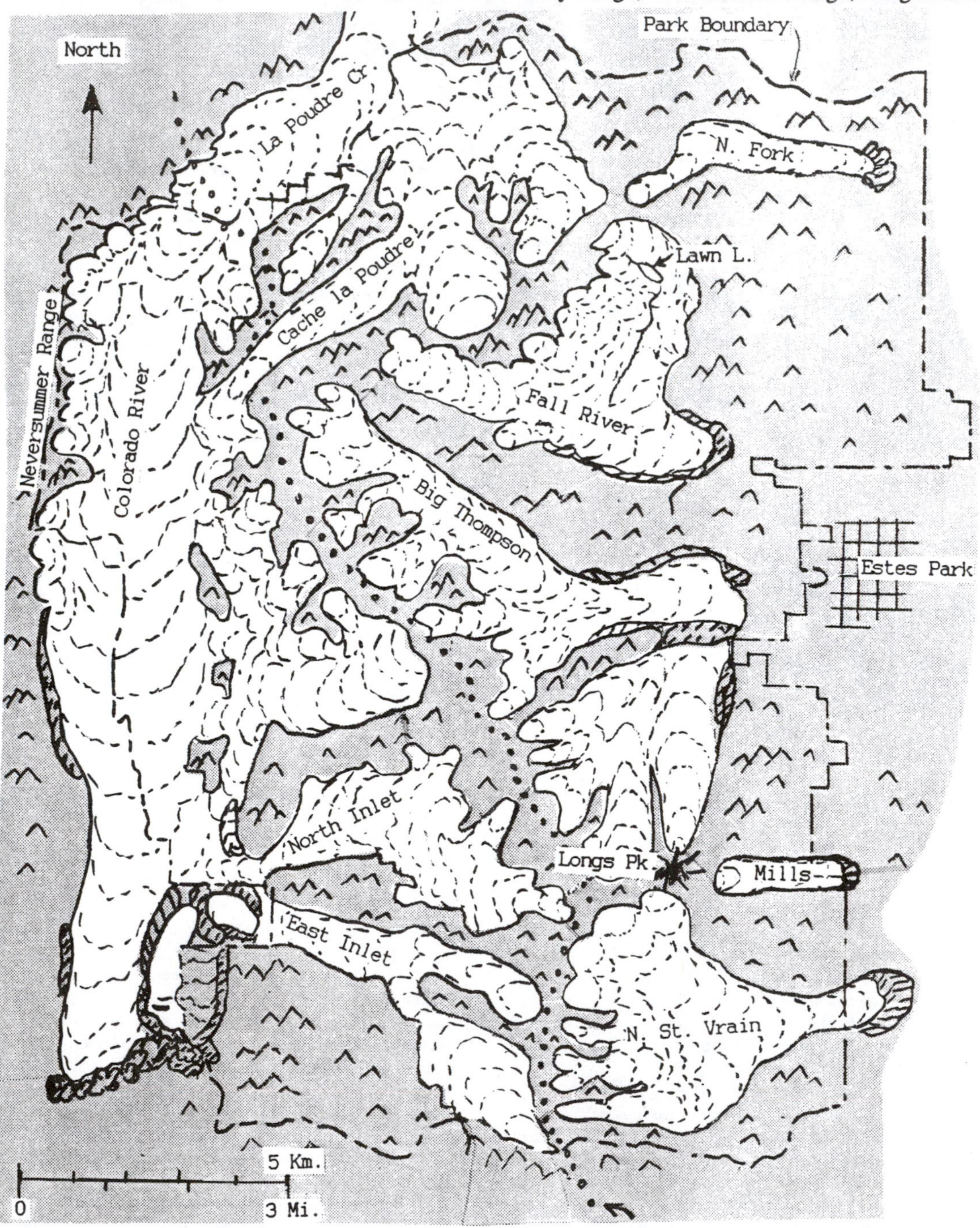

Fig. 62. Map showing glaciers that occupied Rocky Mountain National Park during the Pleistocene.

Peaks are included in the park. It is the largest national park in the state with dimensions of approximately 27 by 40 km.

Obvious features of the park are the world-class exhibits of glacier scour and glacier deposition. Map names like Moraine Park, Bierdstadt Moraine, Taylor Glacier, Cirque Lake, Frozen Lake, Mills Glacier, Glacier Gorge, Andrews Glacier, Iceberg Pass, Tyndall Glacier, Sprague Glacier, Snowdrift Peak and others suggest that the park was the home of a lot of Pleistocene ice. About a third of the park is above timberline (3450 m.) and the National Park Service maps show another 14 unnamed glaciers (or ice fields) within the park and another nine just outside the park boundaries. During the ice age most of the valleys were filled with ice. So much so that the upper portions of the individual valley glaciers coalesced into broad icefields with thicknesses of over 1500 m. in the deeper valleys. Figure 62 is a map showing the extent of the glaciers that occupied the area during the Pleistocene.

All that glaciation made spectacular scenery. Only one "fourteener" is in the park, and that is Longs Peak, at 4276 m. (14,255 ft.) which is, incidentally, the only fourteener north of I-70 in the state of Colorado. Longs Peak is on the crest of a whitish granite stock. Sixteen of the peaks in the park are over 3900 m (13,000 ft.) and nearly 30 are over 3600 m. (12,000 ft.). These peaks are split about evenly between intrusive igneous rocks and Precambrian metamorphics. That mix of metamorphics and igneous masses is typical of the Front Range, and a few other mountain ranges in the state. A few of the intrusive rocks are Tertiary, and the important commercial metal deposits are most often found with the Tertiary intrusives. The thing that makes Rocky Mountain National Park so popular with the tourists is a superb paved highway that goes right over the top at 3655 m. (12,183 ft.). The Trail Ridge Route ascends a broad ridge with a breathtaking view of about 25 km. of the steep north side of the Continental Divide. The north side of the high mountains is the most rugged because the thickest glaciers could accumulate on the shaded side of the mountain. The famous "diamond" on the northeast face of Longs Peak is a 600 m. (2000 ft.) sheer cliff. The normal hiking trail to the top approaches from the west, then attacks the summit on the south face. These peaks all have sheer north faces. A brand of sophisticated backpacking gear is "The North Face" because presumably only the best will work on the north face. Be careful with this idea, though, because south of the equator the south face is normally the steepest.

Fig. 63. Sightseeing in the tundra about 3600 meters above sea level (12,000 ft.).

The Trail Ridge Route crosses some dark volcanic rock near the high point. These are Tertiary, but the source vent is unknown. Specimen Peak in the Neversummer Range is called "The Crater" but it is not volcanic. It does serve as a major breeding area for Colorado's state animal, the Rocky Mountain bighorn sheep.

Lawn Lake is formed by a small earthen dam in the northern part of the park. In 1982 the dam failed and a surge of water and debris roared down Roaring Fork. When the flood reached the floor of Fall River in Horseshoe Park, the debris formed a broad fan that dammed Fall River to form a new lake. Boulders among the debris are over 3 m. in diameter. Horseshoe Park has an unusual display of meanders on the Fall River. The valley was scoured by a massive glacier, and possibly held a moraine-dammed lake when

Fig. 64. Lava cliff at the high point on Trail Ridge Route. These are Tertiary lavas from an unknown vent.

Fig. 65. Jagged ribs of Precambrian metamorphics along the Fall River Road.

Fig. 66. Glacial polish viewed from driver's seat along Fall River Road.

Fig. 67. Boulders from the Lawn Lake flood of 1982.

Fig.68. The Lawn Lake flood of 1982 spread from the left, damming Horseshoe Park.

the glaciers first withdrew from the valley. The flat valley floor allows the present stream to wander aimlessly through the boggy valley. Lulu City is at the very head of the Colorado River. This old mining town was abandoned in the early 1900's after some marginal production of gold and other metals.

Grand Lake is really a *grand* lake. It is the largest natural lake in the state, for one thing, but it has a glacial history that really makes it look grand. The lake is on the west side of the divide and touches the southwest corner of Rocky Mountain National Park. That puts it on the headwaters of the Colorado River. In simple terms, Grand Lake is a moraine-dammed lake at the confluence of North and East Inlet Rivers. These two stream valleys held dandy glaciers that had their sources along the Continental Divide. But the story is more complicated. A moraine complex was formed and the present town of Grand Lake is built on (and among) the moraines. Huge boulders are strewn among the houses. Some are retained for landscaping, but others are just too big to move. The moraine that dams Grand Lake is a terminal moraine from the Inlet rivers, but it also includes the lateral moraine of the Colorado River glacier that was grinding southward from its origin 30 km. away at Poudre Pass. The latest glacial ridge forms the dam for Grand Lake, but with the filling of Shadow Mountain Reservoir the narrow glacial ridge separates the two lakes. Shadow Mountain Reservoir has two more morainal loops near the dam on the south end. These loops would be terminal (or recessional) moraines on the Colorado River glacier. At an elevation of about 2490 m. (8300 ft.) these moraines are typical of the larger glacier deposits in Colorado.

But Grand Lake supplies water to Estes Park, which is on the opposite side of the Continental Divide. There is a 21 km. tunnel from Grand Lake to a small lake at East Portal on Wind Creek, which is tributary to Lake Estes and the Big Thompson River. The extra water helps generate electricity which is returned by cable back through the tunnel and is used to pump water from Lake Granby into Shadow Mountain Reservoir. When Shadow Mountain Lake is full, it backs up into Grand Lake, which provides water to go back down the tunnel. The tunnel is 2.93 m. in diameter with a fall of 33 m., and can carry 12.5 cubic meters (550 cu. ft.) per second when it is full.

Nature must have sensed the need for all that water because in 1976 an intense storm system camped over Estes Park and dumped more than 30 cm. (12 in.) of rainfall in the drainage. With so much bare rock, the runoff was rapid and sufficient to flush the narrow part of Big Thompson Canyon with surge of 10 meters of water. More than 100 persons were lost in the flood.

Fig. 69. Unsorted, unrounded, unstratified till in downtown Grand Lake, Colorado.

Fig. 70. Close-up view of a moraine in Grand Lake, Colorado.

Fig. 72. Krummholz or "banner trees" along the Trail Ridge Route. Icy winter blasts strip all the needles and twigs from the upwind side of the trees. This site is within 200 meters of timberline.

Fig. 71. Moraine boulders that are too expensive to move in downtown Grand Lake, Colorado.

Fig. 73. The lateral moraine for Fall River is outlined in this view from the Trail Ridge Route. Hidden Valley includes the string of beaver ponds along the moraine.

Fig. 74. A granite erratic is part of Bierstadt moraine at the Bear Lake parking lot.

Fig. 75. An erratic rests on a scoured bedrock surface along the trail to Emerald Lake.

Fig. 76. Avalanches pile debris along their "runs" during the winters when heavy snowfalls occur. Most winters get many avalanches.

Fig. 77. Three pronounced morainal ridges are outlined at Moraine Park.

Fig. 78. For the beginners, the Park Service tells where the mountains and moraines are.

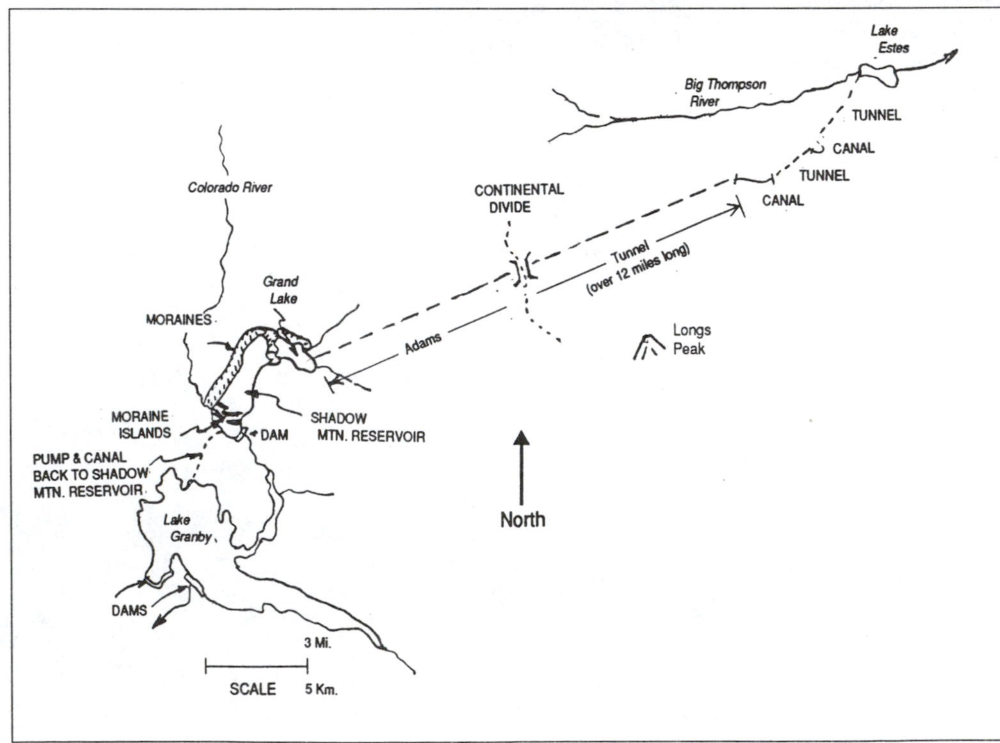

Fig. 79. Water from Lake Granby is pumped to Shadow Mountain Reservoir, which backs water into Grand Lake. The water drops to Estes Park in a tunnel and generates some electricity to pump more water from Lake Granby. The eastern plains keep the water.

Fig. 80. Big Thompson Canyon was scoured by a wall of water when 30 cm. of rain fell in a storm in 1976, killing over 100 persons.

Fig. 81. Highway 34 was rebuilt after the 1976 flood. This new raceway should reduce the damage if a similar flood occurs.

Fig. 82. Pink pegmatite dikes cut mica schist along U.S. 34 in Big Thompson Canyon. With no vegetation in much of the drainage basin, the runoff was very rapid.

Fig. 83. A hiker waves from the steep front of St. Marys Glacier. Summer snowboard enthusiasts often forget the array of boulders at the bottom of the slope. Injuries are frequent.

Saint Marys Glacier

Saint Marys "Glacier" is probably the most famous snowbank for its size in the world. It is probably not true glacier ice and it probably does not move (two criteria needed to be a glacier). However, it does lie at an unusually low elevation to maintain a permanent snow field in Colorado-3300 meters (11,000 ft.). The "glacier" is right at timberline and in a narrow gorge that provides shade most of the year. Hikers can reach the site via about 1.5 km of foot trail at the end of a short dirt road off I-70 near Georgetown.

Fig. 84. St. Marys Glacier nestles in a narrow, shaded gulch above St. Marys Lake. The icefield is not too impressive, but the elevation is barely above timberline.

The Denver Basin

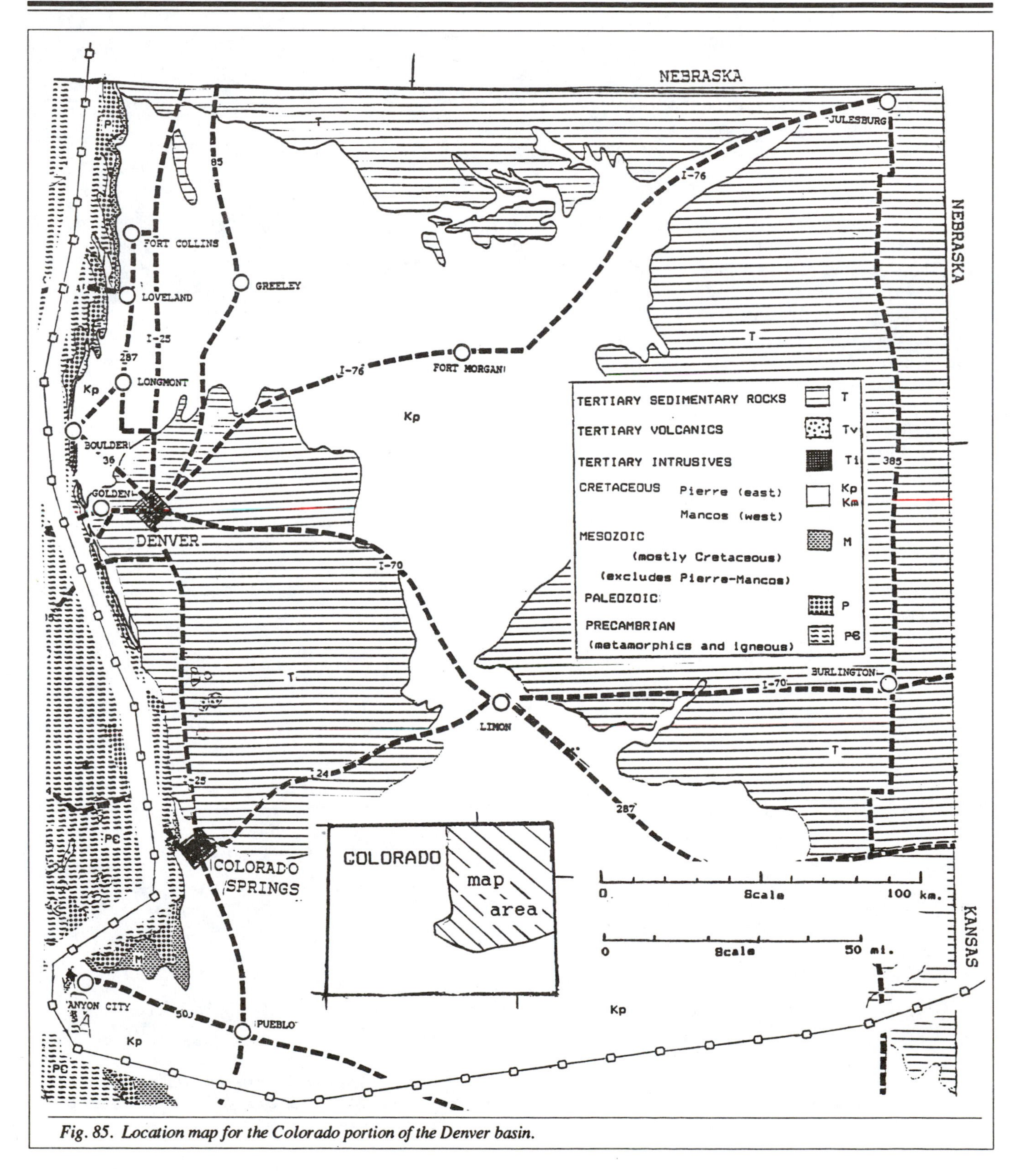

Fig. 85. Location map for the Colorado portion of the Denver basin.

The Denver Basin is a huge syncline that underlies Denver plus most of northeastern Colorado. The Basin begins at about Pueblo and plunges northward into Wyoming, with the gentle east limb stretching into Nebraska and Kansas. A little more than half of the basin is in Colorado, and it comprises about 25% of the total area of the state. At the surface it is very "plain" and a little monotonous, geologically. However, the Denver Basin has great economic value in oil, gas, coal, sand and gravel and a little placer gold in the streams.

Information gained from thousands of holes drilled for water, oil and gas, reveal a broad and very asymmetrical fold. The west limb of the syncline appears as conspicous rows of hogbacks and steep cuesta ridges that extend from south of Colorado Springs, along the Front Range foothills, and into Wyoming. Resistant rock layers are pitched up at 30° to 80° in the typical sequences such as the Pennsylvanian Fountain Formation at Red Rocks Park, near the town of Morrison, and the Boulder flatirons. Near Loveland, the Dakota sands are tilted so steeply that the beds are overturned slightly and they are tilted (dip) to the west at about 80°. The eastern margin of the basin, near the Nebraska-Colorado line flattens to less than 10° of dip. The axis of the syncline is only about 20 km. from the foothills at the base of the Front Range, and the Dakota plunges to more than 900 meters below sea level in that distance.

Fig. 86. The Dakota "hogback" near Loveland is a nearly vertical feature that is overturned and dips about 80 °to the west.

Oil was discovered in 1862 near some oil seeps at Canyon City, which is loosely considered to be in the "basin." The oil at Canyon City is from fractured Pierre Shale, and the discovery was made only three years after the first oil well was drilled in the U.S. at Titusville, Pennsylvania. After more than a century of production, Canyon City and Florence fields still produce marginally economic oil from the Dakota sands. Conoco is a modern oil giant that arose from the early production at Florence.

Fig. 87. A small workover rig repairs an old oil well near Florence. Pikes Peak is directly behind.

Fig. 88. This small oil well pumps from the Dakota Sandstone at a location near Ft. Morgan, at least 240 km. from Florence, but still in the Denver Basin.

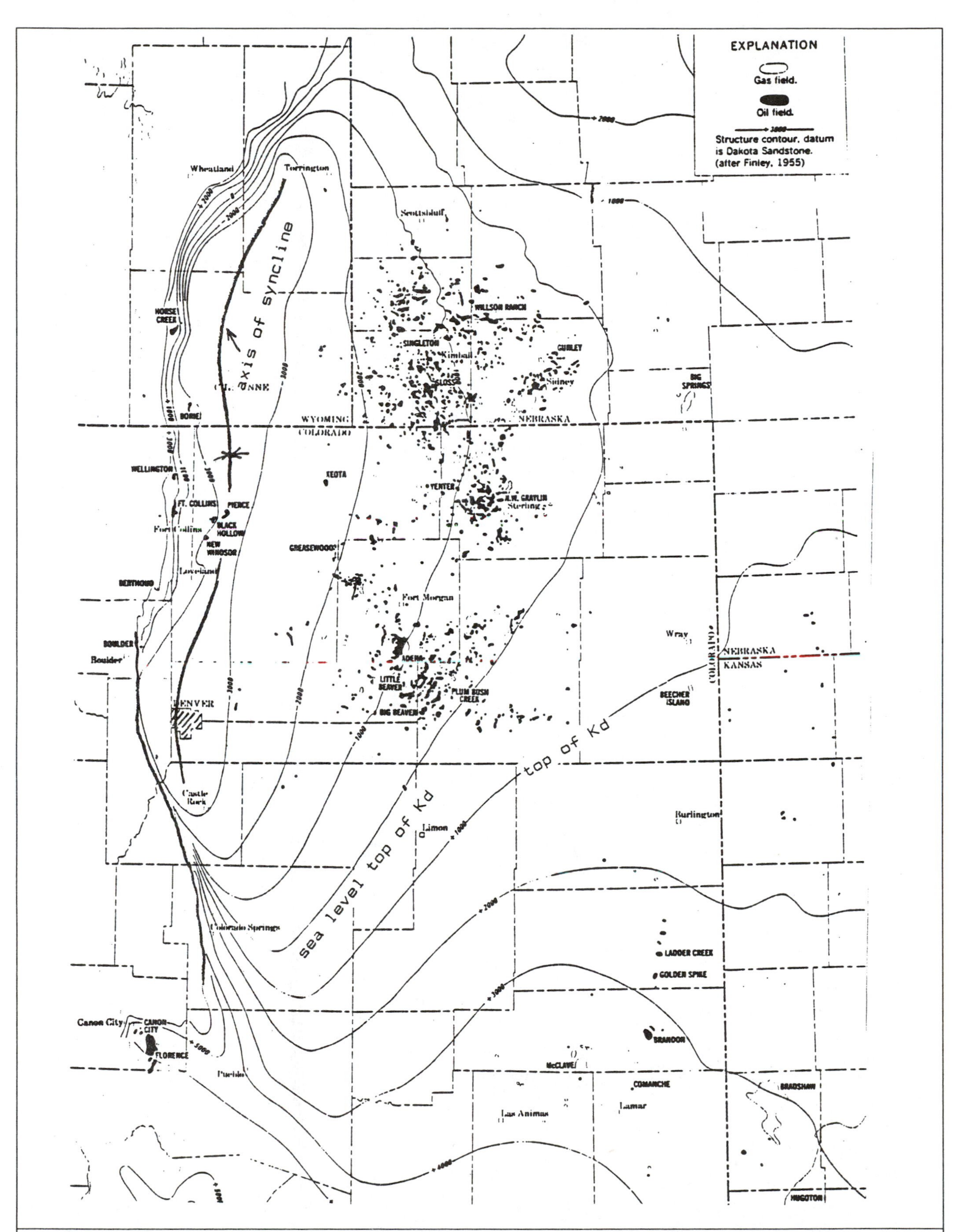

Fig. 89. This structure map of the Denver Basin shows the elevation at the top of the Dakota Sandstone. Note how near the axis of the fold is to the foothills of the Front Range. (From the RMAG Atlas of Rocky Mountain Geology, 1970.)

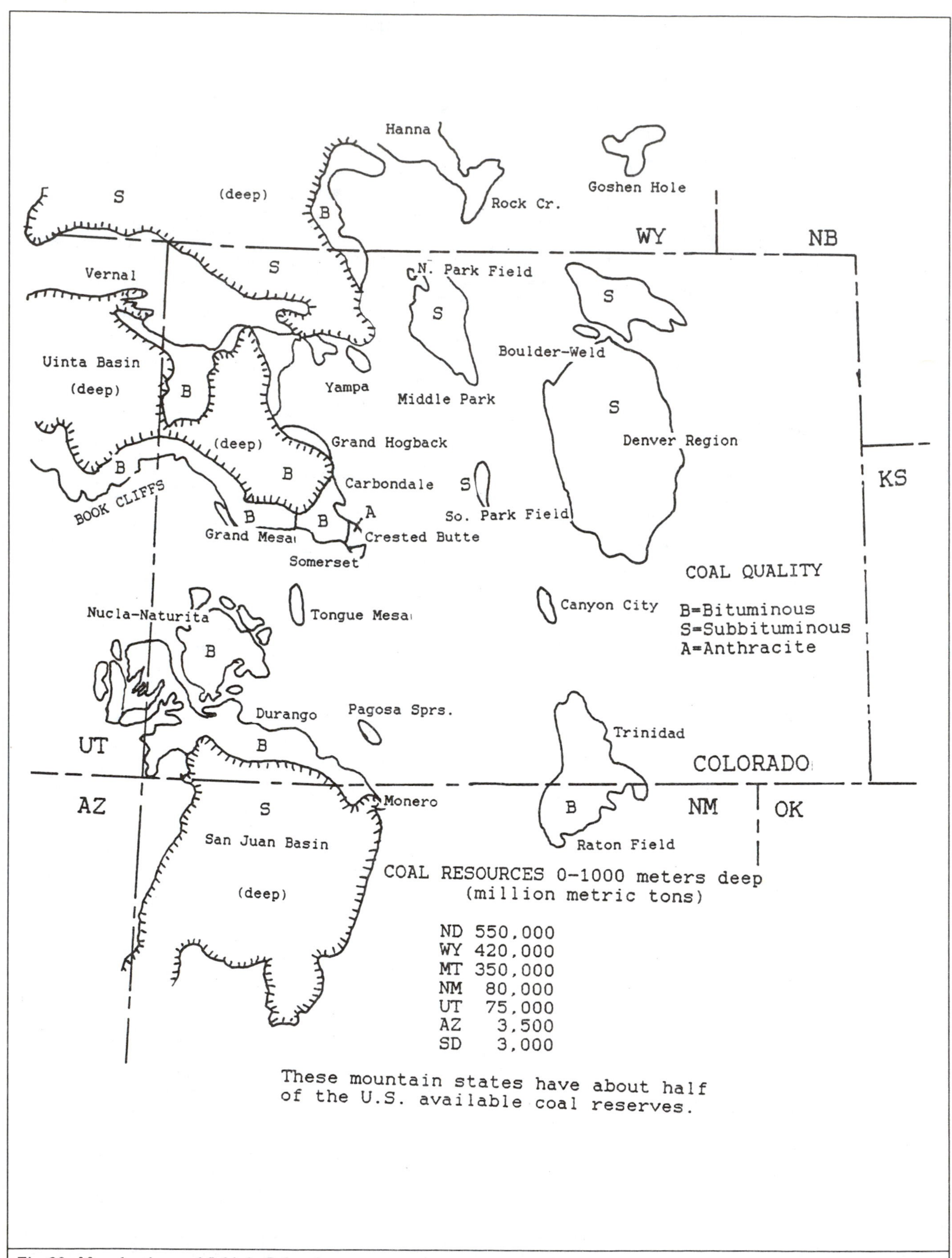

Fig. 90. Map showing coal fields in Colorado.

Fig. 91. Pawnee Buttes are Tertiary remnants that lie about 25 km. south of Nebraska-Wyoming-Colorado corner.

In the Denver basin proper, most of the petroleum (oil and gas) is produced from small sand reservoirs included in the Dakota. Since the 1923 discovery of Wellington Field, near Ft. Collins at 2100 m. below the surface, there have been hundreds of small fields found and the expected production is about 700 million barrels of oil and 800 billion cubic feet of gas. That may sound like a lot of petroleum to a lay person, but Colorado is not even self-sufficient in petroleum. In 1989 Colorado typically produced 78,000 barrels of oil per day, which was only 1.1% of the U.S. production. Colorado's population of three million is about 1.3% of the U.S. total. If foreign imports are included, Colorado supplies only half of the petroleum used in the state. Colorado has one giant oil field in the northwest corner, at Rangely, where about 700 million barrels of oil have already been produced, and the field is nearly depleted. The Denver basin produces approximately as much petroleum as Rangely, but it takes dozens of small fields to make that amount. Many areas of oil and gas production are scattered throughout the state, but Rangely and the Denver basin fields dominate.

Major low-grade coal beds have been mined from the Denver basin and other plains locations. Figure 90 is a map of coal fields around the state. Commercial production in the Denver basin began in Jefferson and Boulder counties in the 1860's but most of the coal in the plains is of poor grade, and nearly all the modern production is from higher grade bituminous beds along the New Mexico boundary and from the western counties. The early production near Denver was prompted more by the location near large populations rather than quality of the coal. Industrial coal requires a better quality than most home-heating uses. Coal in the Denver basin is latest Cretaceous. The mines are generally shallow, although some coal has been taken from more than 300 meters deep.

Near the northeast corner of the state, just south of the Wyoming-Nebraska-Colorado intersection, there is a distinctive blemish in the otherwise smooth complexion of the plains. At Pawnee Buttes the Tertiary formations have not been stripped completely from the older rock layers. If these buttes occurred anywhere west of Denver, they would not be worth mentioning, but in the flatlands of the high plains, they are rather significant.

Fig. 92. Castle Rock Conglomerate, Oligocene (?) exposed south of Denver at the town of Castle Rock.

Castle Rock is a small table-like butte (small mesa?) about 50 km. south of Denver. Again, it is a small feature where the Castle Rock Conglomerate forms a protective cap for some softer Tertiary sediments that underlie it. In places the conglomerate has some coarse volcanic boulders in it, but it is less than 30 meters thick and of rather limited distribution. If it did not have a town named for it few people would notice the unit. The U.S. Geological Survey Lexicon gives the Castle Rock an Oligocene age. Castle Rock is shown in Fig. 92.

Another small, Tertiary remnant that has had all the surrounding material removed, is the Monument at the town of Monument. When viewed from the east in the morning, the early sun illuminates the thin rib so that it resembles the Washington Monument. Figures 93 and 94 are not from the correct angle, but the comparison with the Washington Monument is obvious.

Fig. 93. Monument, Colorado is named from the spire that looks, from the east end view, like the Washington Monument.

Fig. 94. The monument at Monument, Colorado is a remnant of Tertiary sedimentary rocks that once covered this area near the foothills, 65 km. south of Denver.

The Spanish Peaks Area

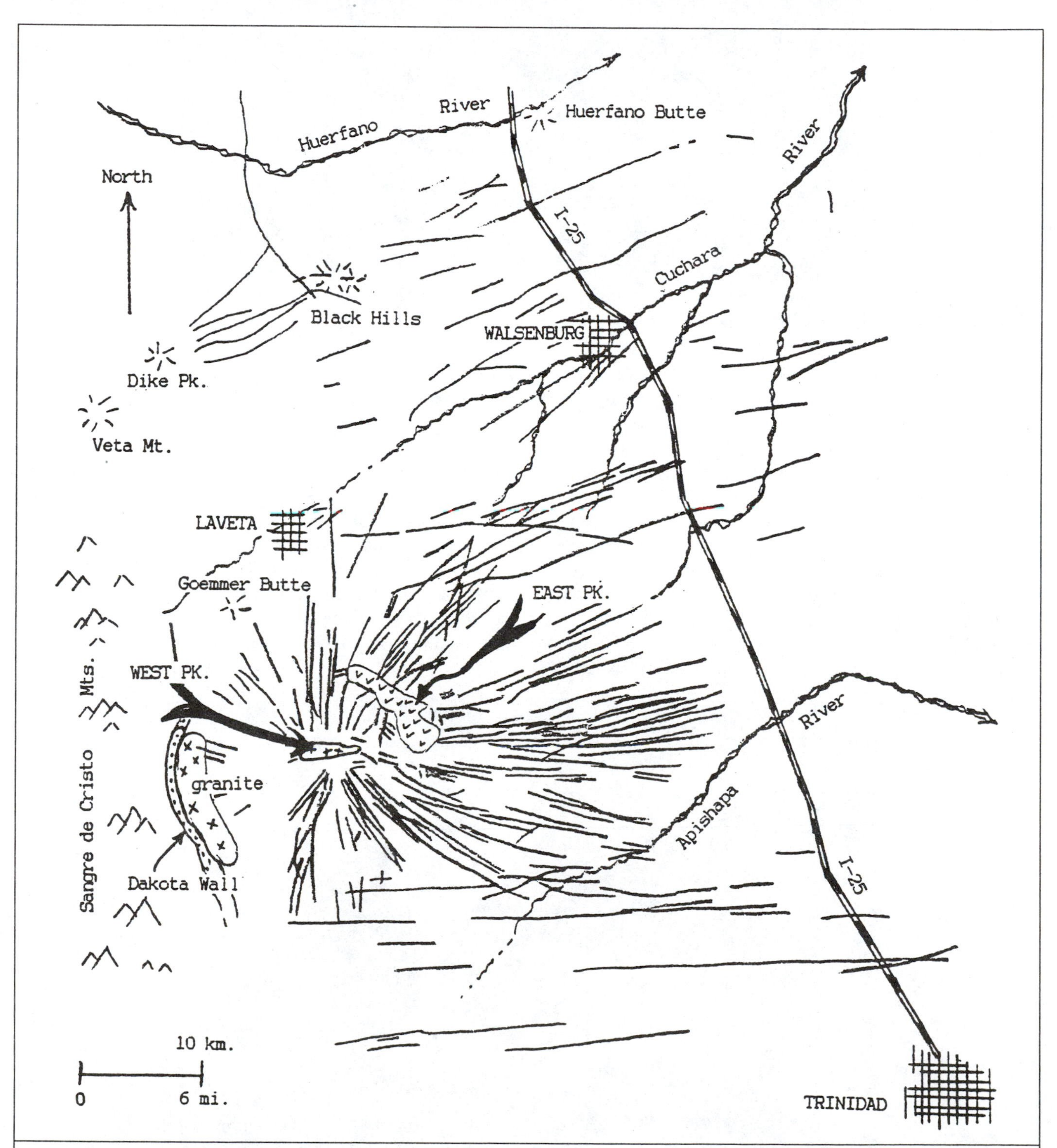

Fig. 95. This map is modified from one by Adolph Knoph in a 1936 bulletin from the Geological Society of America. The swarm of dikes is a world-class feature.

Fig. 96. Soft Pierre shales weather away from hard, intrusive dikes that radiate from the Spanish Peaks.

The Spanish Peaks, just 40 km. north of the New Mexico state line are known to geologists the world over for a spectacular system of dikes that radiate from them. These two stocks extend farther into the plains than does Pikes Peak, and they are not genetically related to the Sangre de Cristo range 15 km. to the west. The stocks push up through the Pierre Shale and they are very fine-grained intrusive diorite (?). As the magma punched its way up-ward, some of the molten material filled the fractures that spread outward. There are dozens of dikes, ranging from a few centimeters in width to more than 10 meters, and extending more than 8 km. across the margins of the plains and the Rocky Mountains. Soft shales of the Pierre have eroded away from the dikes, leaving them as awesome walls. West Spanish Peak, the higher of the two, stands at 4,080 m. (13,600 ft.).

To confuse the issue, not all the vertical walls in the area are intrusive dikes. At the Dakota Wall Ranch, on Highway 12, near the town of Cuchara, the Dakota sandstone got caught between the Spanish Peaks and the east margin of the Sangre de Cristo mountains. As if to mock the igneous dikes, the Dakota forms a similar wall. However, the joints tend to be parallel to the sedimentary bedding in the Dakota wall, and the joints in the dikes tend to be normal to the tabular dikes.

Fig. 97. The massive dikes near the Spanish Peaks tend to have horizontal, columnar joints.

Fig. 98. Windows form as the columnar jointing in a dike breaks down.

Fig. 99. The Dakota wall resembles an igneous dike, but the jointing is mostly along the sedimentary stratification layers. A closer look shows the rock to be sandstone.

Fig. 100. The mailbox identifies the "Dakota Wall Ranch" with the wall in the background.

Fig. 101. Huerfano Butte is an "orphaned" knob of very dense, black gabbro.

Huerfano Butte

"Huerfano" means orphan in Spanish. If you study a highway map of Colorado there is a town of Huerfano, the Huerfano River and Huerfano County, but the cause of all the huerfanos is so small that it does not occur on most highway maps. About 15 km. north of Walsenburg, and in full view of a turn-out on U.S. I-25 is Huerfano Butte. This isolated knob of intrusive igneous rock pokes up through the Pierre Shale and rises about 60 meters above the average surface of the surrounding plains. The black rock is so hard that fresh blocks ring like metal when struck with a hammer. Boulders at the margin exhibit spheroidal weathering, either because they have been subject to a longer period of weathering, or the molten rock along the margin picked up more contamination from the shales of the Pierre host rocks. The Butte is a simple, weathered-out volcanic neck or plug, and if the plug ever reached the surface, that evidence is gone along with the hundreds or even thousands of meters of sedimentary rock that once covered the area. The Huerfano is about the smallest item to be described in this book, but it is so much an orphan to this part of the plains it deserves mention.

Fig. 102. Marginal parts of the Huerfano exhibit spheroidal weathering.

The San Luis Valley

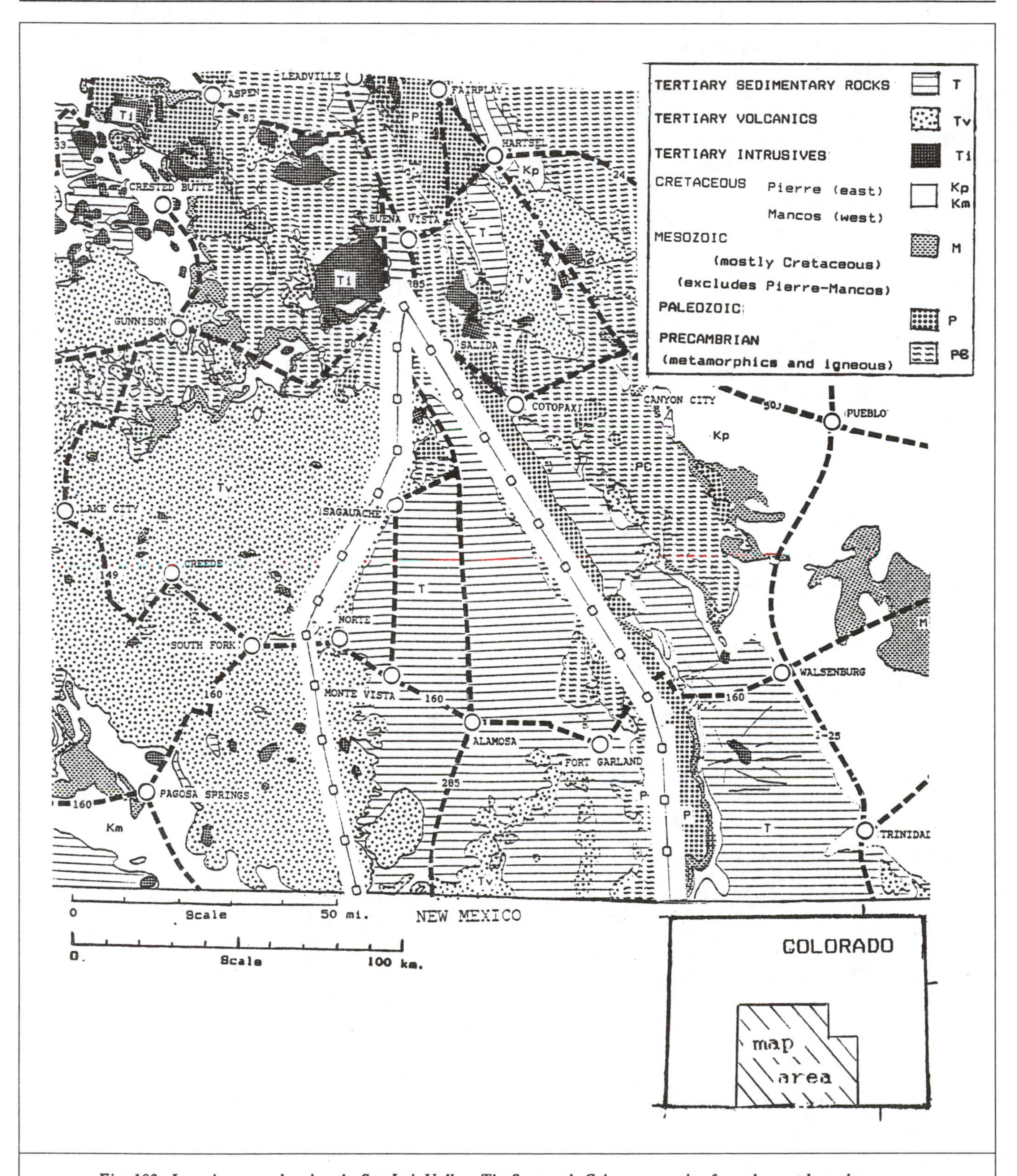

Fig. 103. Location map showing the San Luis Valley. The Sangre de Cristo mountains form the east boundary.

The San Luis Valley is a large, and nearly flat, fault-bounded trough, that stretches about 160 km. north-to-south, with a maximum width of about 100 km. About 10,000 sq. km. of the San Luis Valley are in Colorado, which is about 4% of the entire area of the state. The valley floor is about 2,750 m. (7,500 ft.) above sea level and when clear winter skies cover the area, Alamosa is often the coldest town reporting in the contiguous states. The cold temperatures result from heavy and cold, night time air, that sinks into the basin from the surrounding very high mountains. If there is snow cover on the ground, the cold temperatures are exaggerated by temperature inversions. In an inversion, the heavy and cold air on the ground does not mix with warmer air above, and each night even colder air drains off the nearby peaks, making the condition worse. When the snow in the valley melts, the daytime sunshine will eventually warm the surface and break the inversion. Because of the cool and short growing season, crops in the San Luis Valley are limited to hay, potatoes and other short-term plants, even though there is plenty of water and the soil is good except in the salty bottomlands where the drainage is poor.

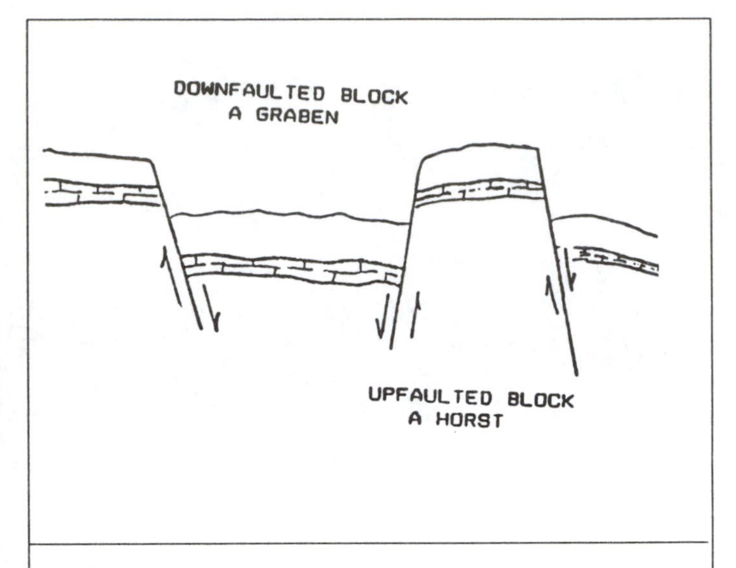

Fig. 104. A simplified sketch of a graben and horst.

When a trench-like feature is formed by a down-dropped block of crustal rocks, we call the feature a GRABEN (German for trench). The San Luis Valley is even more unique. It is a rift valley. A rift valley is a tear in the earth's crust, where a continental plate is actually splitting apart and one side is "drifting" away from the other. The rift of the San Luis Valley has separated to the north probably as far as Steamboat Springs, and extends southward, under a lot of volcanic material, far into New Mexico. The San Luis Valley was recognized as a rift only recently, but it is similar, geologically, to the mid-ocean rift in the center of the Atlantic Ocean, and to the great Rift Valley of Africa. When crustal plates separate, the rift is typically bounded by normal faults (to make the trench) and an upwelling of basaltic lava along the fault lines. In the San Luis Valley, basalts are abundant on the west side of the valley, and basalts are interlayered with gravels under the valley.

The San Luis Valley is a classical "ARTESIAN" basin. When underground water is under a confining layer of impervious rock, and there is abnormal pressure on the water, we say the water is ARTESIAN. For example, if a farmer drills a water well and encounters water in a confined layer at 100

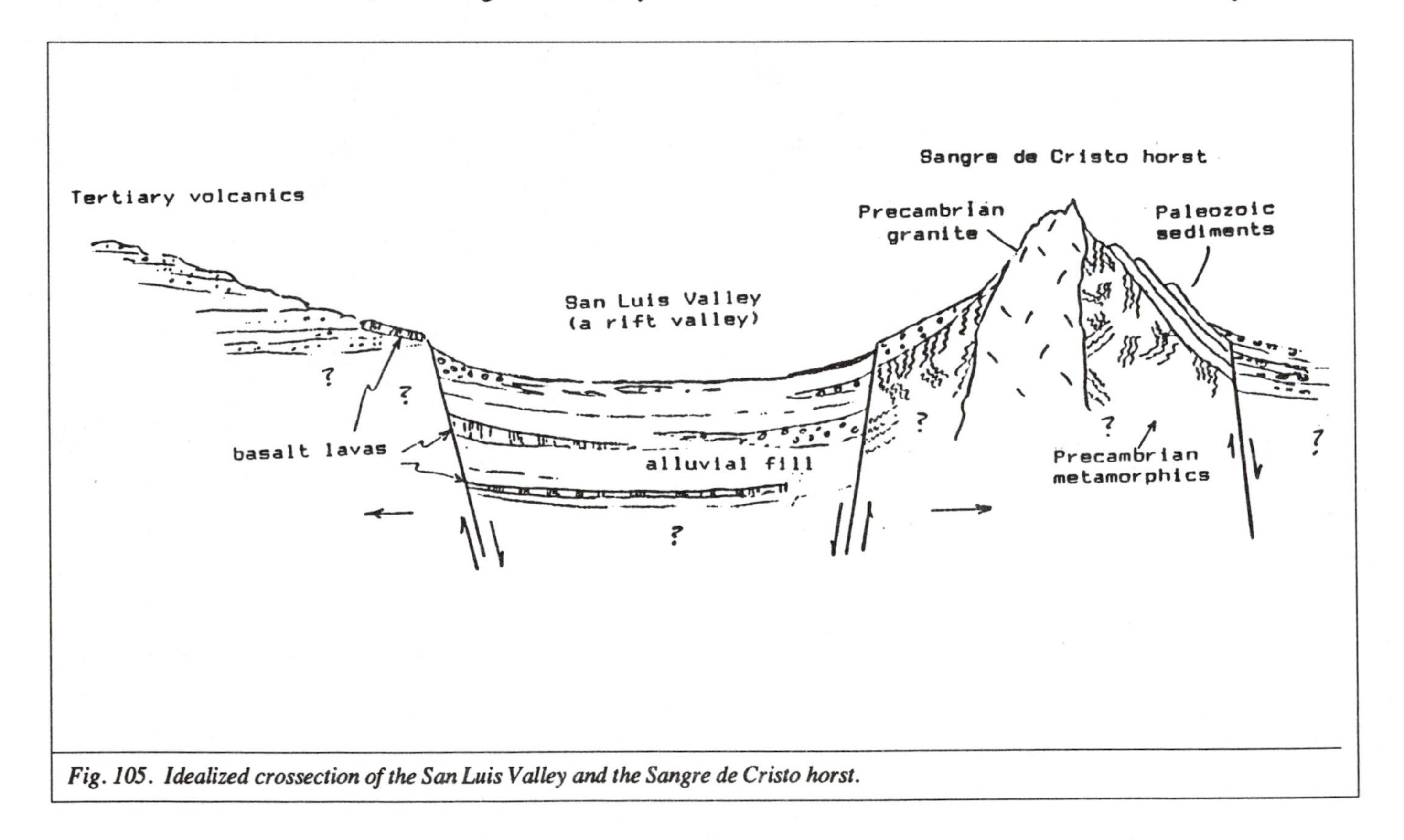

Fig. 105. Idealized crossection of the San Luis Valley and the Sangre de Cristo horst.

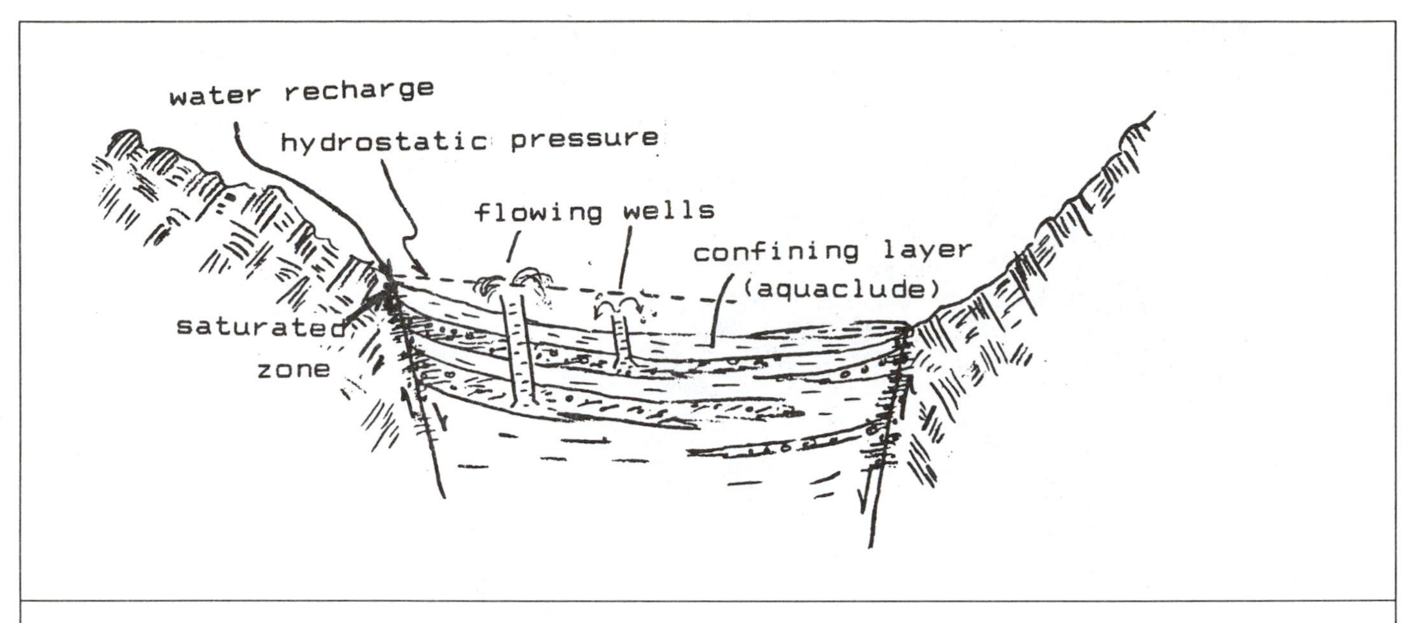

Fig. 106. A sketch showing an artesian water system similar to the San Luis Valley.

Fig. 107. Approaching the San Luis Valley from the west, multiple basalt flows line the valleys. Here three distinct sequences of flows form their characteristic "terraces."

Fig. 108. Some of the flows penetrate the flat floor of the San Luis Valley. Lake sediments fill the basin because the south end of the broad valley is frequently blocked by material that spreads from numerous volcanoes in New Mexico. Several tributaries to the Rio Grande converge in the valley.

Fig. 109. Indian Head is a small volcanic neck at Del Norte.

Fig. 110. Blanca Peak, fourth highest in the state, is composed of light-colored Precambrian granite. Here a skiff of snow adds the white. Soil and water in the San Luis Valley are excellent, but the growing season is very short.

Fig. 111. Powerful pumps lift artesian water from the basin which is then sprayed from these crawling sprinklers. Water pressure helps to drive the devices, which can water a 40 -acre circle with only one setting.

meters below the surface, but the water immediately rises in the well bore to some level above the level it was first encountered, it would be artesian water. The well would be an artesian well. If there is enough pressure to force the water to the surface, the well would be a FLOWING ARTESIAN WELL. Some artesian systems have so much confining pressure that the water squirts out like a powerful fountain at the surface.

In the San Luis basin, gravels, sand and finer sediments are spread across the valley from streams that discharge along the mountain fronts on either side. If a porous layer of sand and gravel is covered with clay and mud, the porous layer can be confined by the clay. As the valley subsides in the fault trough, and the sediments continue to fill the basin, many sequences of water-bearing "AQUIFERS" have formed. Lava flows have also spilled into the valley, complicating the artesian systems. The part of the valley that lies in New Mexico even has a few cinder cones in the valley, and from time to time the southward drainage of the Rio Grande River has been dammed by volcanic events, causing the valley to become a temporary lake. When the lake spills over, the loose volcanic material is soon removed, allowing the basin to dry out once again. There are thousands of feet of sediment in this ever-deepening trench.

The Great Sand Dunes National Monument

Fig. 112. This interpretive sign at the Great San Dunes National Monument explains how the sand is moved to and then dumped at the site of the dune field.

At the east side of the San Luis Valley, the prevailing westerly winds have piled up a cluster of sand dunes that reach 210 m. above the valley floor. Winds that would be strong enough to move the sand grains farther are forced to lift the grains over a stream named Medano Creek, then maintain the carrying power through the foothill brush and trees, then rise over a group of fourteeners. The shape of the Sangre de Cristo range has a small cove where the sand tends to accumulate.

The dunes are composed of sand from a variety of rock types, including feldspars and darker minerals from both the metamorphic rocks to the east and volcanics to the west. Hikers on the dunes are tempted to take off their shoes to cross the cool sands of Medano Creek (often a damp sand-filled wash). If they leave the shoes behind and then climb the dunes, bright sunshine on the dark sand will heat the surface sufficient to cause second degree burns to the feet.

Fig. 113. The Great Sand Dunes snuggle against the east side of the San Luis Valley, at the foot of the Sangre de Cristo Mountains.

Fig. 114. The dunes rise more than 210 m. above the valley floor, making them "great" indeed. The dominant wind blows from the left (west) in this view and Fig. 113, and any sand that tries to blow out of the valley falls into Medano Creek, in the foreground, and is returned to the west end of the dune field (a natural environmental program!).

Zebulon Pike entered the valley during the winter of 1807, after leaving some frostbitten members of his scouting party in a valley to the east of the Sangre de Cristo range. Pike noted the impressive dunes in his journal, then entered the main valley, only to be captured by the Mexican Army. Captain Pike explained that he was looking for a different valley, but as a U.S. Army detachment "invading" Mexican territory, he was hauled off to Santa Fe, seriously delaying his return to his frosty troops.

The Sangre de Cristo Mountains

The Sangre de Cristo Mountains (Spanish for "blood of Christ") are not a very large part of Colorado, but they are a majestic range of mountains. With only one percent of the area of the state, this 2,800 sq. km. lump has nine fourteeners, including the fourth highest Blanca Peak, one of the more spectacular summits of Colorado, at 4,304 m. (14,345 ft.).

Most of the Sangre de Cristo range is a HORST; that is a range uplifted and the two flanking valleys down-dropped by normal faults. The broad San Luis Valley on the west is part of a rift valley where the crust is separating and the resultant tension, or stretching of the crust has allowed the San Luis block to drop down. Basaltic lavas have oozed up around the west side of the San Luis block but so far they are not obvious on the east side, against the Sangre de Cristos. East of the Sangre de Cristo range is the Wet Valley, which is a narrow graben, with generally untouched scenery that belies the fact that the rustic old town of Silver Cliff is less than 100 air kilometers from Pueblo, Colorado Springs and Fort Carson.

Geologically the Sangre de Cristos are a smashed pile of Precambrian metamorphics that were heaved up during the Laramide Orogeny and dragged some Paleozoic sedimentary rock units along with them. The eastern flanks of the range have several thousand meters of Mississippian, Pennsylvanian and Permian sediments exposed, but the summit areas are all Precambrian crystalline rocks. Blanca Peak is part of a small Precambrian stock, and this nearly white peak is aptly named "Blanca," which is Spanish for "white." Most of the year it is white because it is snowcapped, and the rest of the year the granite stands out in contrast to the more somber colors of the gneisses and schists of the Precambrian. It also pokes up more than 1,000 m. above timberline.

The Upper Arkansas River Valley

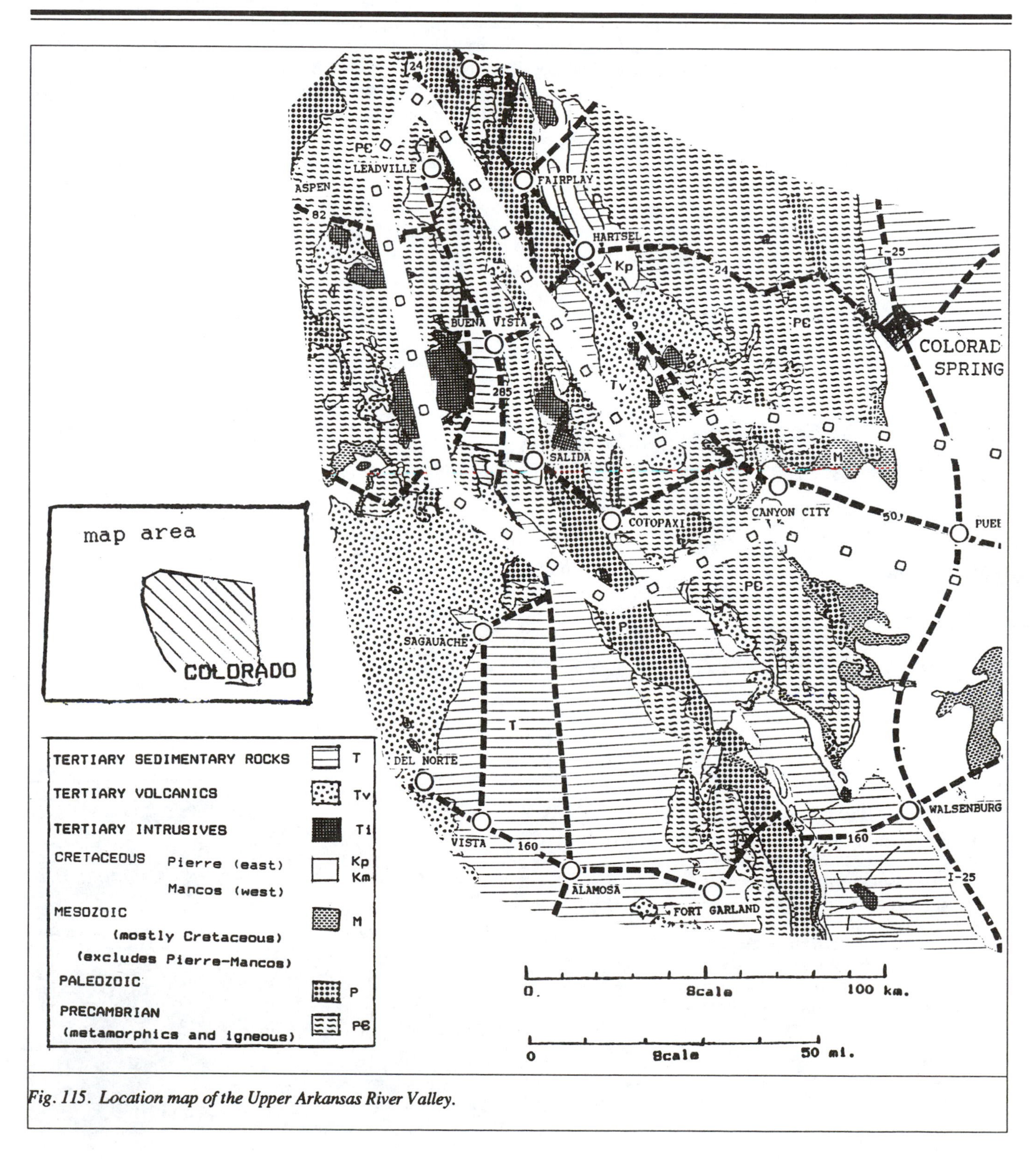

Fig. 115. Location map of the Upper Arkansas River Valley.

The Arkansas River begins under Mt. Elbert, the highest fourteener in the state at 4330 m. (14,433 ft.). After flowing southward past Leadville, Buena Vista and 18 fourteeners (5 on the east side of the Arkansas valley and 13, including the five Collegiate peaks on the west side), it reaches Salida, where it breaks out of the graben valley which comprises the upper 100 km. of its journey. The graben valley of the upper Arkansas is a continuation of the rift that is responsible for San Luis Valley. The Arkansas once was tributary to the Rio Grande. At Poncha Springs there is a thick pile of volcanics, called the 39-mile Volcanic Field, and apparently in the late Tertiary, lava flows dammed the upper Arkansas drainage and impounded a deep lake. When the lake spilled, it went out the east side, across the northern tip of the Sangre de Cristo range.

Today, if you want to climb over the dam in the graben, the road leads over Poncha Pass, which is 2700 m. above sea level. The top of the dam, therefore, is 600 m. (1970 ft.) above the present river level at Salida. The dam did not necessarily make a lake 600 m. deep because there has been much downcutting since the river was diverted. Once it spilled into the plains, the river had a very steep gradient and easily chewed its channel downward. In the early stages there was a lot of soft, alluvial material, but eventually the loose material and the soft sedimentary rocks were removed and the river hit the hard, Precambrian basement rock. Its ability to cut was not hampered much, however, as attested by the colorful canyon between Salida and Coaldale. At Coaldale the Arkansas makes a right angle turn to the northeast and heads straight for another gorge at Canyon City.

Back in 1929 engineers built a bridge across the spectacular gorge at Canyon City. At more than 310 m. high, it is a remarkable structure, and is still advertised as the highest suspension bridge in the world. But it doesn't lead the tourist any place important. Because the bridge is there, visitors go to the Royal Gorge Park. If you cross the bridge, it only leads to the other part of the park—and you return across the bridge. So the bridge is really a monument to bridge-building. It does provide an unmatched view of the gorge. From the Royal Gorge the Arkansas River runs unimpeded 240 km to Kansas and on to the Gulf of Mexico via the Mississippi River.

Fig. 116. The north suburbs of Leadville nestle among large mine waste piles. Although placer gold along the streams lured the early prospectors into the area, Leadville is one of the leading silver producers in the nation. Galena Peak (galena is lead sulfide) is one of the towering peaks to the east of towm.

Fig. 117. Turquoise Lake is a moraine-dammed lake immediately west of Leadville. During part of the Pleistocene, Leadville was probably covered with ice, as the elevation of the town is 3060 m. (10,200 ft.). The late stages, including the Wisconsin, left large ridges of the morainal till alongside and at the end of the lobes that extend down the various valleys. A small manmade dam was built to plug the part of the moraine loop that was breached by the stream.

Fig. 118. Twin Lakes are about 16 km. south of Leadville where Lake Creek enters the Arkansas River, and Route 82 joins U.S. 24 along the Arkansas River. Route 82 crosses the Continental Divide at Independence Pass (3630 m.) just west of Twin Lakes and proceeds down the colorful Roaring Fork Valley to Aspen. Mt. Elbert, the state's highest peak is only about 10 km northwest of Twin Lakes. Note the recessional moraine in the middle that makes the lake a "twin" lake.

Fig. 119. Several surges are shown in the lateral moraine on the north side of Twin Lakes. Mt. Elbert is in the background.

Fig .120. Huge erratic boulders are among the till that forms the moraine dam at the downstream end of Twin Lakes.

Fig. 121. Placer gold miners have screened the sand out of the moraine and outwash terraces on the downstream end of Twin Lakes.

A

Fig. 122. Place names along the upper Arkansas give a hint of the local geology. Gold Camp Trailer Park (A) and the Placer Bar (B) would suggest that there is gold "in them thar hills." The town of Granite (C) calls attention to a small granite stock that is somewhat atypical of the predominantly metamorphic rocks in the region.

B

C

Fig. 123. Obvious till is exposed in a small morainal ridge that is cut by U.S. 24, near the town of Granite.

Fig. 124. A lateral moraine dwarfs a house near Granite. The house rests on outwash material that was flushed from the glaciers as they retreated from the valley.

Fig. 125. Fourteener Mt. Princeton is the high point of the Princeton granite stock, and is the southernmost of the Collegiate Peaks. Princeton Hot Springs bubble up on the bleached, south (left) flank of the peak. If you remember that feldspars weather to clay minerals (and chalk), you should expect that someone would name the creek Chalk Creek.

Fig. 126. Chalky Princeton granite exposed along Chalk Creek at the Princeton Hot Springs. Fifteen kilometers up Chalk Creek is the beautifully preserved, rustic ghost town of St. Elmo.

Fig. 127. The Mosquito Range forms the east boundary of the Arkansas River Valley. If one looks eastward over the Mosquito Range, the 40 km. by 80 km. South Park can be seen. South Park is drained by the South Platte River, and the treeless basin is over 2700m above sea level.

Fig. 128. Antero Reservoir is on the west end of South Park. The Mosquito Range rises in the background. Much gravel fills the structural basin, and the north part, near the town of Fairplay, has been screened by huge dredges to remove placer gold.

Fig. 129. About 25 km. north of Leadville, Route 91 crosses the Continental Divide (3400 m.) at the town of Divide and the molybdenum mining complex of Climax. The tailings pond of the world's top molybdenum production occupies this basin, just below timberline.

Fig. 130. Heading west from Leadville, and after crossing Independence Pass, Route 82 winds its way down the headwaters of the Roaring Fork toward Aspen.

Fig 131. The upper Roaring Fork emerges from winter in early July. The rocks are mostly Precambrian metamorphics, but several granite stocks can be seen along Route 82 near Aspen. Glaciers scraped the U-shaped finishing touch to the valley here.

Middle Park - Eagle Valley

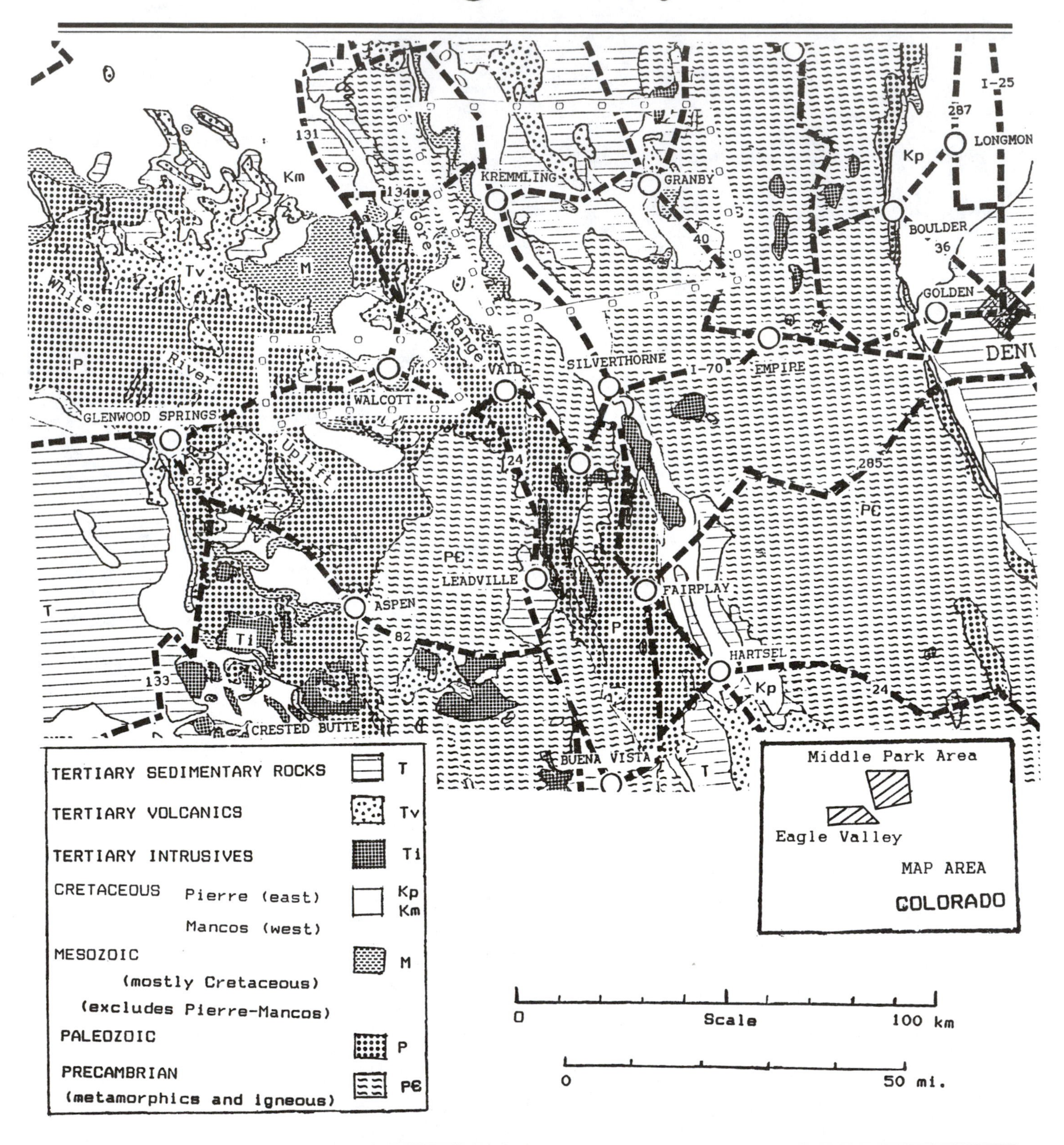

Fig. 132. Location map of the Middle Park and Eagle Valley areas of Colorado.

This section on Middle Park- Eagle Valley is a catch-all section, and it is a great place to introduce some of the material about valuable metal deposits. Colorado is the world's leader in molybdenum production, and has a colorful history of gold, silver, lead, zinc and copper production. Cripple Creek gold deposits have been mentioned in a previous section, but now it is time to explain how the metal got into the mountains. Cripple Creek gold and most of the ores in the San Juan Mountains were introduced from volcanic pipes that vented at the surface. Later phases of the volcanics may have behaved like a stock, but the following generic variety of mineral deposit associated with a stock fits most of the important mining districts of the Western U.S., Canada, Mexico, and South America.

Figure 133 is a sketch of a mineralized stock that has melted and pushed its way toward the surface among several types of sedimentary rocks. As the stock enters the area, it folds and cracks the adjacent beds, fracturing the brittle formations and maybe only folding the clay shales and mudstones. Wiggly lines indicate a border zone that has been metamorphosed by heat and reactive, hot fluids from the molten stock. Mineral vapors and hot fluids have penetrated the area. These vapors and fluids may have contained iron, lead, zinc, gold, silver, copper molybdenum or any variety of valuable metals. Lead, silver, copper, gold, iron and zinc are commonly found together in the metal deposits compounded with sulfur (galena is a lead sulfide, chalcopyrite is copper-iron sulfide). The enriched fluids permeate the host rock until the correct temperature/pressure/chemical conditions force the metals that are in solution to precipitate as mineral crystals. When tin, tungsten or molybdenum are present, they precipitate at relatively high temperatures, and mercury and antimony are precipatated at relatively cool temperatures. The other sulfides overlap a great deal in intermediate deposits. Sometimes the rocks and minerals seem to ignore these basic rules, making the search for valuable ores very challenging and often frustrating.

Fig. 133 shows the rich metal veins as dark cracks radiating

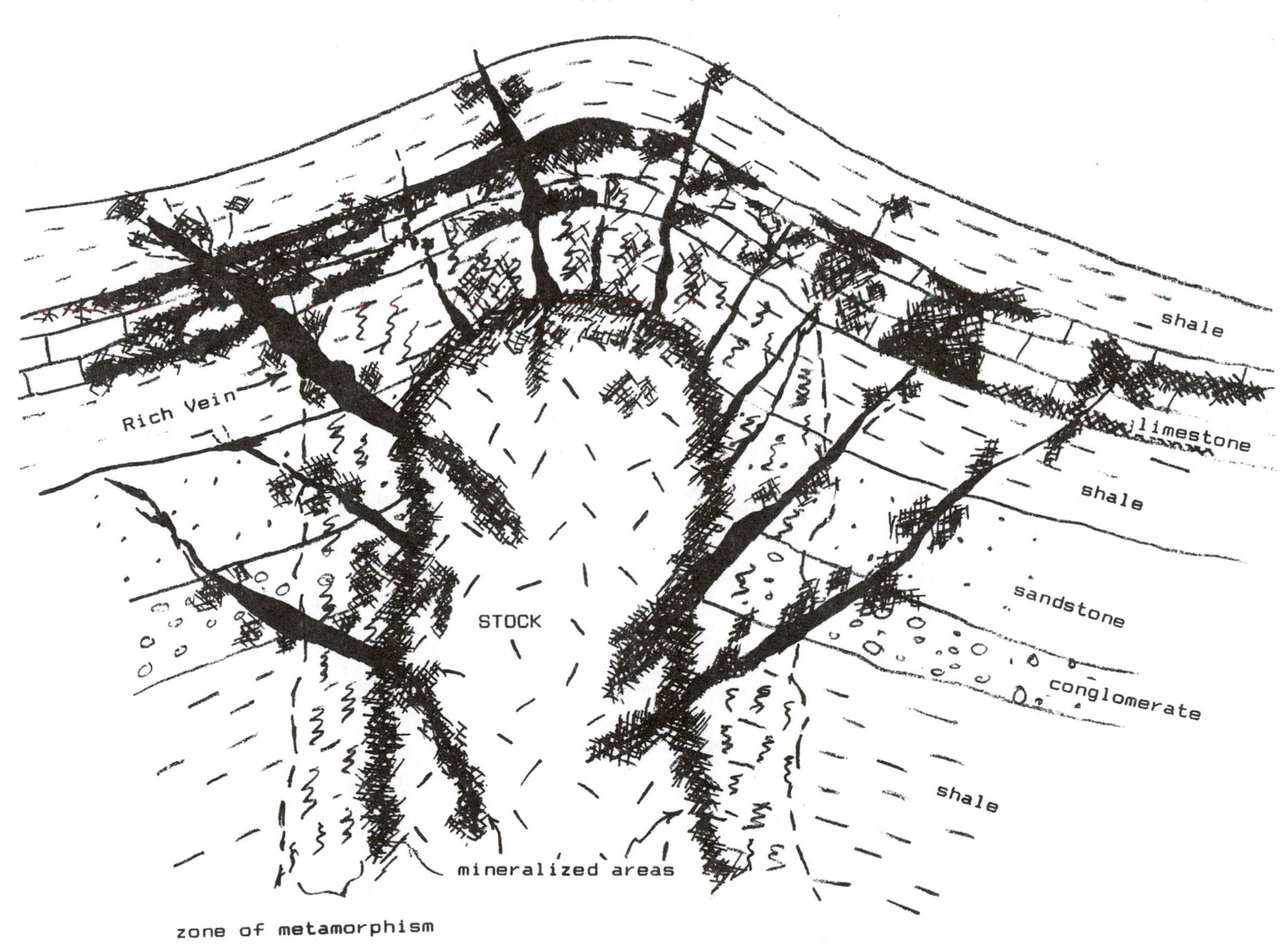

Fig. 133. Idealized sketch of an igneous stock that has pushed and melted its way into a sequence of sedimentary rocks. Note the contact metamorphic zone around the stock, the dark veins of minerals and the shaded areas of partial mineralization. Limestones are favored sites for mineral exchanges with the host rock.

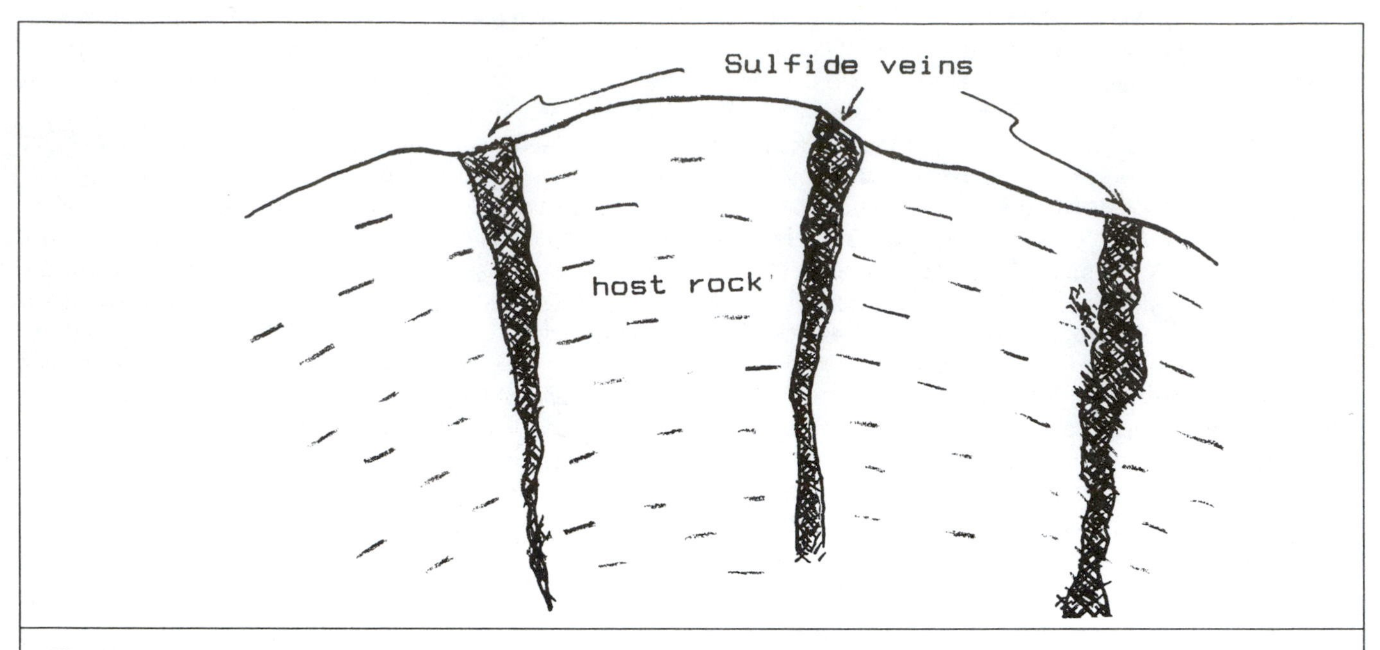

Fig. 134. Sketch of three mineralized veins exposed at the surface.

from the stock. Lesser mineralization is shown as shaded areas, and can occur in the surrounding rock, near cracks, and in contact with the central stock. When the stock cools, it shrinks somewhat, allowing mineralization within the stock. Note the rather wholesale mineral replacement in parts of the limestone. Limestone is by far the favored host rock for chemical exchanges of valuable metals.

Iron is an abundant metal. If any valuable material is present, there is nearly always abundant iron - enough so that if a prospector searches only for the iron minerals, he can expect

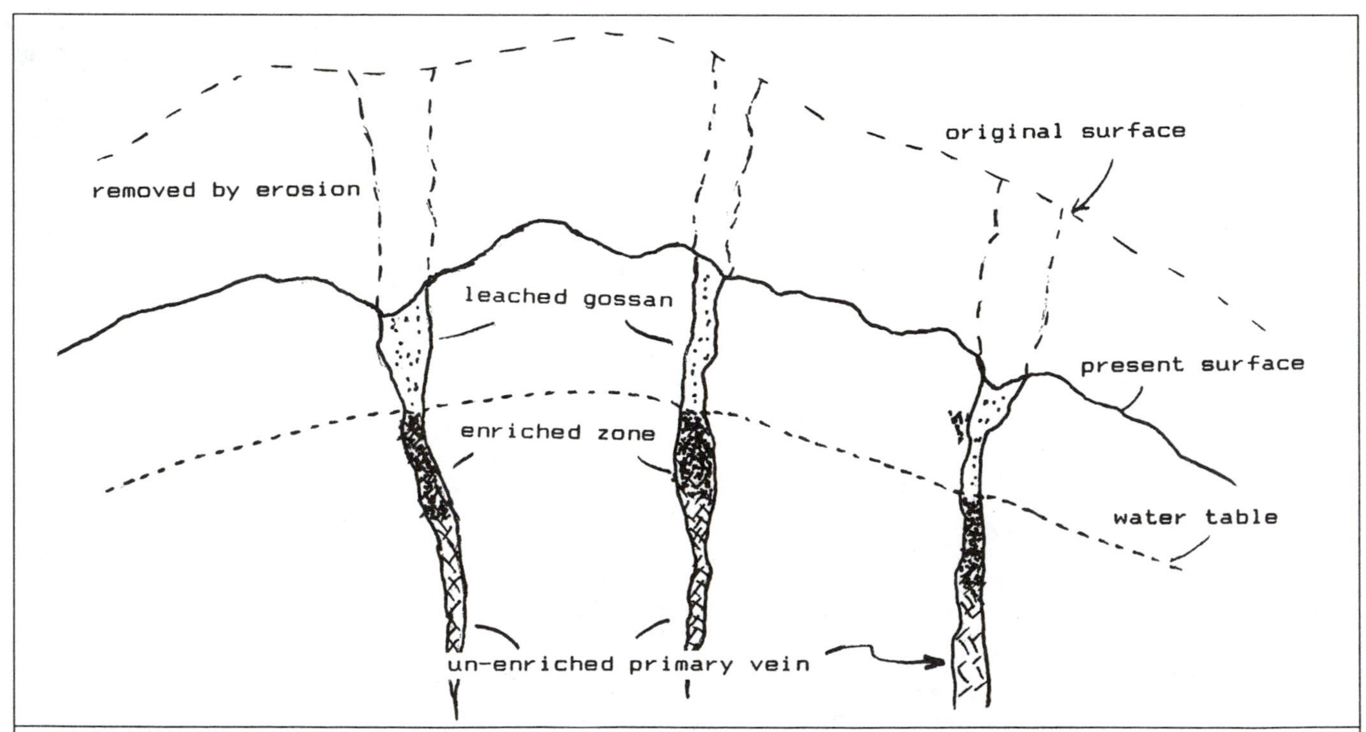

Fig. 135. Sketch of mineralized veins after much of the area has been removed by erosion. Veins have been leached by downward percolating, acidic ground water, leaving a limonite-rich "GOSSAN." In many veins, only coarse gold will remain after the sulfide ores (copper, zinc, lead and silver) have been leached away. Often the sulfides are reprecipitated in an enriched zone at the water table.

to find all of the valuable veins by just finding the iron-rich deposits. Iron sulfide (pyrite, or fool's gold) is the common iron sulfide. When it weathers, it converts to iron oxide (limonite) and causes the soil to become richly colored with yellow and brown "rust." Figure 134 shows veins of rich sulfide material at the surface, before being weathered. Let's assume that the ore minerals include Fe, Cu, Pb, Zn, Ag and Au, and that the richness of the veins would be worth $50 a tonne.

In Figure 135, the vein has been exposed for many thousands of years and some of the surface material has been leached out by percolating groundwater and the valuable metals have migrated downward, in solution, to the WATER TABLE. As erosion progresses, the water table continues to move downward. When the percolating minerals reach the saturated water table, the chemistry of the water may be different enough to cause the percolating solutions to re-precipitate a new suite of minerals-in addition to the original sulfide minerals that have not been leached away. Enriched zones often occur at the water table, and they may be several tens of meters thick, grading downward to the un-enriched, primary sulfide ore vein. The enriched zone may be worth $100, $500 or even $10,000 per tonne. The un-enriched zone below the enriched material may still have the original $50 per tonne values, and the overlying leached zone may be worthless. These crumbly, brown, leached mineral veins are called GOSSANS. Gold is very stable, chemically, and if significant gold were present, especially if it were larger then microscopic specks, the gold may remain in the leached zone. Prospectors quickly learned to check the limonite-rich leached zones for gold. If high gold values remained, they would mine the vein down to the water table to seek the enriched zone. If the vein were barren in coarse gold, the vein often was bypassed, leaving undiscovered rich veins of Pb, Zn, Ag, Cu and fine Au. If gold remained in the leached vein, it could be stripped by erosion and added to the streams that drain the area. These little pieces of gold would be washed downstream and become PLACER GOLD as part of the stream sands and gravel. Each springtime flood would wash the gold farther downstream and allow the pebbles to pound and mash the grains to flatter flakes, such as shown in Figure 143. Nearly all streams that flow out of Colorado contain some placer gold among the gravels.

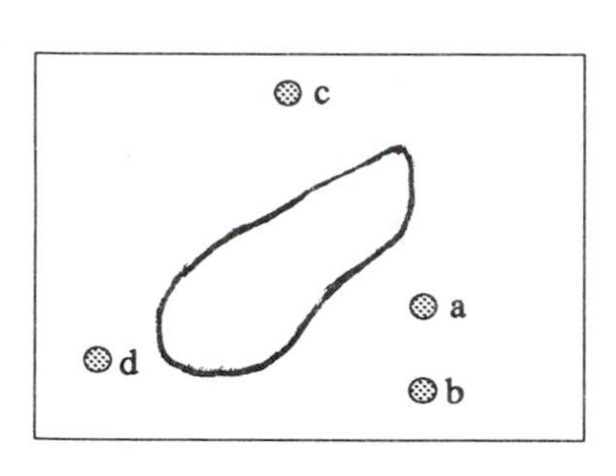

Fig. 136. The so-called "Colorado Mineral Belt" is a zone stretching across the middle of Colorado that includes most of the Tertiary metal deposits that brought mining into the state in the late 1800's. Site (a) is the amazing Cripple Creek, which is not in the belt, and the Spanish Peaks (b) and Hahns Peak (c) have good deposits that are outside of the belt. Site (d) is Uravan which is the center of a uranium-vanadium mineralized "belt."

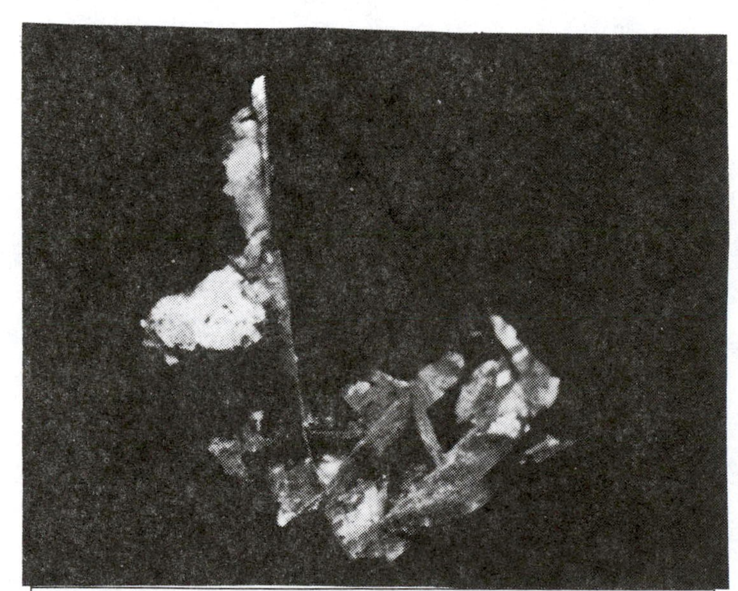

Fig. 137. This chunk of pure gold, on display in the Denver Museum of Natural History, was extracted from a lode deposit by strong acids. The interconnected network of gold shoots was freed by removing the surrounding host rock. Tiny gold veinlets are often linked together. The display in the museum includes a large collection of these acid-freed chunks. This one is about 15 cm. across.

Many articles about Colorado show the "Colorado Mineral Belt" that runs through the center of the state and includes most of the great old mining camps. Telluride would be on the southwest end, and, moving northeastward would include Silverton, Ouray, Lake City, Creede, Powderhorn, Ruby Mountains, Gunnison Basin, Aspen, Minturn, Arkansas River, Fairplay, Breckenridge, Leadville, Climax, Silver Plume, Idaho Springs, Central City, and Blackhawk, to name just a few of the old camps. The actual concept of a mineral belt has some flaws because some good mines have been worked at Hahns Peak, north of Steamboat Springs, Figure 136 (c), and at the Spanish Peaks (b). Cripple Creek (a), was probably the most exciting camp in the state in its heyday, but it falls outside of the "mineral belt" of most authors. Uravan (d) was the center of the "Uravan Mineral Belt," but it contains the important uranium-vanadium deposits that did not get much attention until about 1945, after the "mineral belt" had been cast in stone.

Mineral veins encased in the bedrock are known as LODE deposits, as opposed to PLACER deposits that occur when the valuable minerals are included with gravels in the streams that drain the lode areas. Because gold is the most famous of placer minerals in Colorado, we usually forget that platinum, diamonds, tin, and a variety of other materials are recovered from stream and beach gravels around the world. Figure 137 is a gold specimen that is exhibited in the Denver Museum of Natural History. The museum has a special area to show the tremendous variety of minerals collected in Colorado. The "gold room" is awesome. Colorado School of Mines has a valuable display also, including many Colorado specimens, but also spectacular material sent in by alumni of the school, which include some of the world's top mining engineers who have managed some of the classic world mines for over a century.

Fig. 138. The old zinc mining town of Gilpin lies abandoned along U.S. 24 between Leadville and Vail. The hoist was in the white building on the top terrace and mine wastes made the terraces on the right.

Fig. 139. From Gilpin the mine tails were dumped over the steep slope extending down to the Eagle River. Retaining walls were built to keep the material from the river, but after a century of aging, the walls have mostly failed.

The gold specimen in Figure 137 has been extracted from a rich lode vein by dissolving the vein with strong acid. Gold is almost inert in the acids that will dissolve limestone, feldspar and even quartz. The museum has a case full of huge networks of these etched-out specimens. They show that the gold fills tiny fractures in the hard rock, often interconnected over several centimeters.

Lode, or "hardrock" mines pock-mark many of the rugged mountains and hills of Colorado, especially within the so-called mineral belt. Figures 138 and 139 show 1991 views of the old mining town of Gilman, which lies on a hillside between Leadville and Vail, along U.S. Highway 24. Gilman was a zinc and lead mining center, and the waste and tailings from the big operations were dumped on the hillsides according to the standard procedures of the Old West. As pollution from these old historic operations seeped into the streams, it became obvious to most observers that some sort of cleanup was needed. Gilman, and the smelter site on the Eagle River at Minturn became the focus of many environmental outcries, because the beautiful Eagle River was so loaded with percolating iron sulfates and other, more toxic solutions, that the boulders along the stream were brightly coated with limonite. Minturn was gone by 1990 and the tailings and much of the fouled soil had been removed and hydro-mulched for restoration in 1991. Mining companies who wish to reopen the old diggings or start a new mining venture in Colorado or any other state now, must clean up the abandoned sites and put restoration money into a trust fund to clean up the operation when the mining is completed. In fact, the rules are so tight in Colorado that most mining companies would rather move to Nevada or other states that are a bit more hospitable to mining. Nevada

Fig. 140. A view of waste placer gravels from dredges at Breckenridge. This 1986 scene now includes part of an airport and a new shopping center. Without new topsoil, vegetation will not reclaim these waste piles for centuries. Modern mining laws require restoration money held in trust before a mine can begin altering the countryside.

Fig. 141. The author is shown scooping gravel into a coarse screen, resting on a gold pan. Using a rock pick to loosen the gravel, it is easy to collect a pan of sand and gravel. The sand is screened into the pan, and the gold is isolated in the pan by swishing and pouring out the lightweight grains. Gold is approximately 10 times as heavy, under water, as normal rock grains. Gold trails down the streams that drain the gold lodes so that there are still tons of gold in Colorado streams, but the concentration is too low to be commercial in all but a few sites. (Photo by Jack Berry)

Fig. 142. A raft drifts past an old river terrace near the town of Radium, about 20 km. downstream from Kremmling. Dashes show the layers of bedrock in the Morrison Formation that underlie the abandoned gravel bed of the Pleistocene river channel. These tilted hard and soft layers made a natural riffle that should help concentrate gold at the bottom of the former stream channel. The circle indicates the location of an old placer gold mine that operated prior to 1950. Most of the gold came from the Breckenridge area, at least 120 km. upstream.

now leads the nation in gold production. However, lest you blame it all on bureaucracies, Nevada has a lot of fine grained gold that is suitable for cyanide heap-leach extraction.

Most of the big lode gold deposits were found after placer gold was found in the streams and the prospectors followed the "colors" up to the outcrop. Leadville, Blackhawk, Fairplay, Idaho Springs, Telluride, Rico, Silverton, Creede, Lake City and other camps grew from the tiny gold clues in the streams. The town of Auraria was a tent city inside what is now Denver in 1858 when prospectors found gold in the stream there. And some prospectors were on their way home from California where they had gone broke in the big 1849 California gold rush. They had learned how to pan for gold in the streams, and they found enough in Clear Creek and other forks of the Platte River to linger and search in Colorado. In the winter of 1858-59 the lodes were found at both Central City and Idaho Springs, and the discoverers of both lodes had to flee the winter storms and return in the spring. Imagine their anxiety as they huddled in the cold snow of Auraria during January and February while they waited for the snow to melt—knowing all the time where the "mother lode" was for two major gold fields.

Figure 140 shows gravel piles near Breckenridge, at the headwaters of the Blue River, which is a major tributary to the Colorado River. Similar piles of coarse gravel surround Fairplay, and they appear along the upper Arkansas River, Clear Creek, the Dolores River and many other gold-bearing streams. Large floating dredges scooped the gravel from the valleys and the fine gold was sorted out before the waste gravel was discarded from a boom extended from the side of the dredge. After World War II the dredges were phased out because they absolutely devastated the pristine meadows of the mountain valley streams.

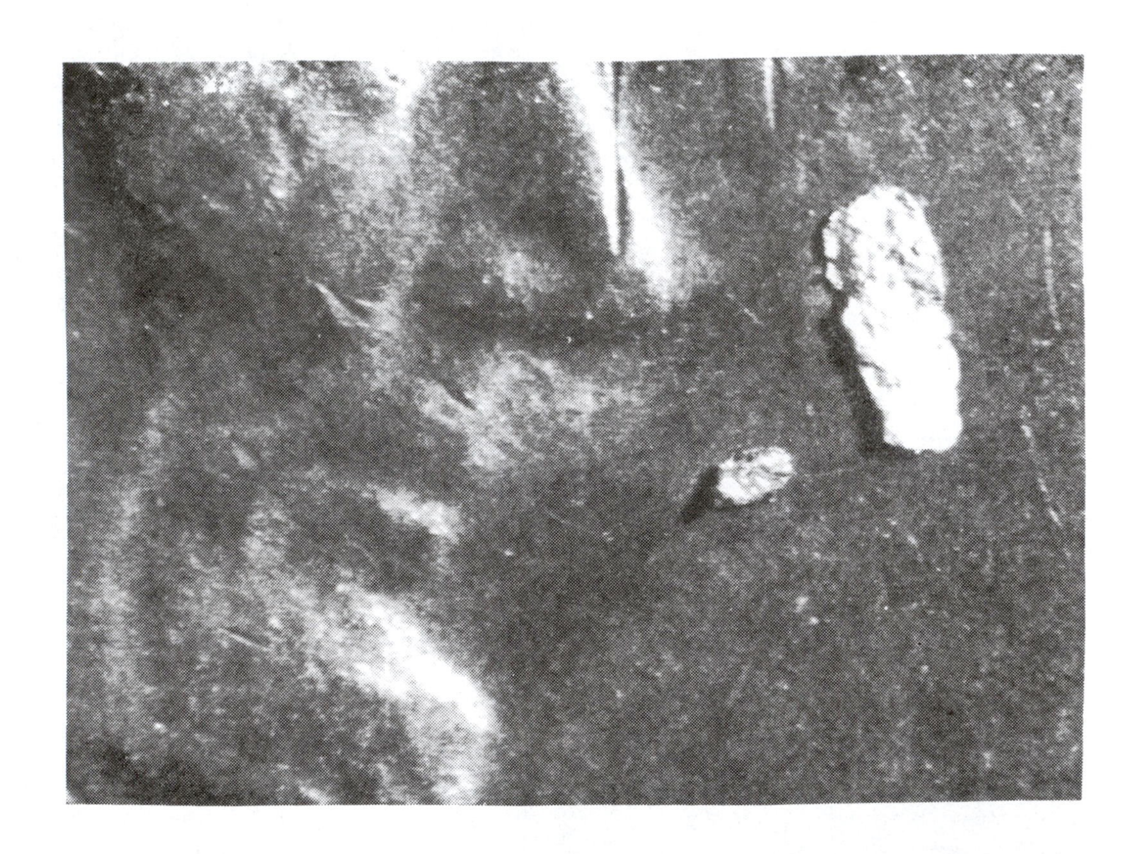

Fig. 143. Two tiny gold flakes are shown on a battered U.S. penny for scale. The larger flake is about 1.5 mm. long, and in 1990 was worth about as much as the penny. (Photo by Michael Eatough)

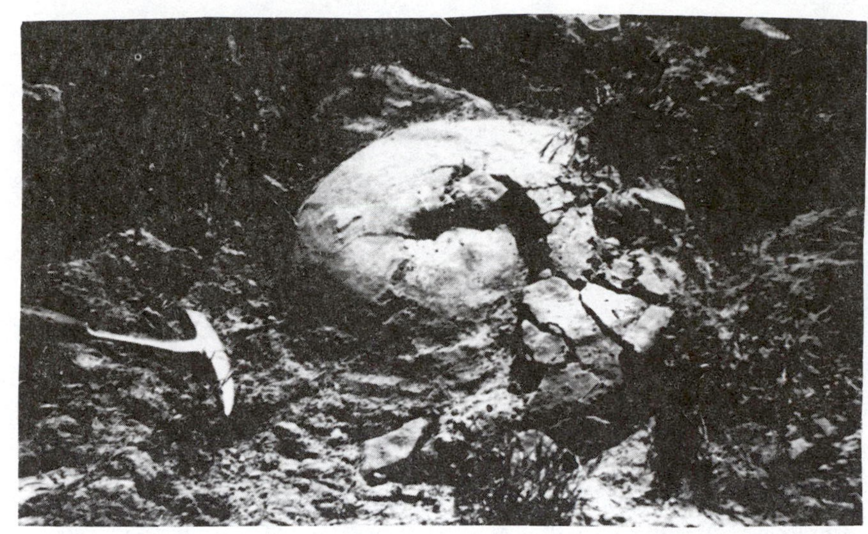

Fig. 144. A large specimen of the fossil cephalopod Placenticeras as uncovered from the Mancos-Pierre at the "Bird Bath locality." The pick head is about 17 cm. long (7.5 in.).

Fig. 145. The outside mold of a Placenticeras shell. With the fossil gone, the depression holds rainwater and forms a perfect bird bath, hence the name "Bird Bath Locality." A long pencil marks the scale.

Fig. 146. It is appropriate for a woman named Robin Kibler to use a bird bath for washing.

Fig. 147. This view of the Bird Bath locality shows a sandy zone of the Pierre-Mancos that forms a resistant ridge. The fossils are contained in concretions in the sandy zone. Hundreds of the fossils have been weathered out and extracted by collectors.

At Breckenridge an airport and several housing developments have been built on the old dredge discards. It will take centuries to get a good vegetative cover on the piles.

Kremmling is a small town on the Colorado at the confluence of the upper Colorado River and the Blue River, that drains the rich mining districts around Dillon, Silverthorne and Breckenridge. The Colorado wanders toward Kremmling through a subdued park-like terrain from Lake Granby, crossing a lot of soft shale. About 20 km north of Kremmling is an interesting outcrop of the upper Mancos-Pierre Formation known as the Bird Bath locality. There is a sandy sequence in the shale that contains a great number of the large cephalopod fossils called *Placenticeras*. Figure 144 shows one of the large critters, exposed in the loose soil. Most of the fossils are encased in a boulder of sandstone, and the early collectors of the fossils usually split the boulders with a large sledge and opened the two slabs to remove the fossil. With the fossil animal gone, the two halves of the encasing boulder each would form a crude basin, that can collect rainwater. The locality has an unusual assemblage of fossils, and the U.S.B.L.M. has withdrawn the area from collecting, except for special permits issued for scientific research.

After the Colorado River reaches Kremmling, and joins the Blue River, it plunges into the Gore Range. As with the Front Range, and many other ranges in Colorado, this Tertiary upfold has warped and wrinkled some younger beds around the core of the mountain, but the core is composed of hard, Precambrian metamorphic rock. Igneous dikes and sills may streak the central core of the hard rock. In the case of Gore Canyon, downstream from Kremmling, the railroad actually follows the

Fig. 148. Dave Wolny, a museum researcher, examines a cluster of bird bath concretions. The fossils are gone from most of the exposed boulders.

Fig. 149. A cluster of the large clam Inoceramus lies exposed at the Bird Bath Locality. Inoceramus is a familiar genus in the Mancos and Pierre Formations and their equivalents over several states.

Fig. 150. After drifting through the poorly defined "Middle Park," the Colorado River plunges through the Precambrian core of the Gore Range, just west of Kremmling. The D&RG Railroad follows the whitewater canyon through the gorge, but the highways avoid it. This view is from a road cut on the south side of the gorge, showing the rail line below. Maximum relief in the gorge is about 600 meters.

Fig. 151. Along the "Colorado River Road" between State Bridge and Dotsero, the Pennsylvanian Eagle Valley Formation is exposed, showing much gypsum and thin-bedded siltstones and shale. At Sweetwater Canyon, a few kilometers from the Colorado River, the Eagle Valley Formation is tightly folded into a recumbent anticline. The force to flop these beds over came from the east, during the Laramide Orogeny.

Fig. 152. The White River Uplift raises a large chunk of Colorado to over 3200 meters between State Bridge and Meeker. One of the most remote portions of the uplift is the Flattops Wilderness Area. Trappers Lake is a beautiful gem at the entrance to the wilderness area at the headwaters of the White River at 3000 meters above sea level. The lake is surrounded by thick flows of basaltic lava, making the mountains "flat topped."

Fig. 153. Eagle Valley is carved by the Eagle River from soft, folded gypsum and salt layers of the Eagle Valley Formation, of Pennsylvanian age. This view, looking west near the town of Gypsum, shows the sharp anticlinal fold that is presumed to be a squeezed-up salt dome, now breached by erosion.

Fig. 154. These contorted, gypsum-rich beds of the Eagle Valley Formation are seen from the westbound lane of I-70 at the Gypsum exit.

river down the canyon (Figure 150). Below State Bridge, the Colorado River enters an area of thick, gypsum-rich Pennsylvanian beds known as the Eagle Valley Formation, with the type section at Eagle, Colorado. The Eagle Valley Formation was deposited in the Maroon Basin at the same time the Paradox Formation was deposited in the 4-Corners area (Paradox Basin, of Pennsylvanian age).

The east end of the Eagle Valley is marked by Red Canyon, where bright red beds of the late Pennsylvanian Maroon Formation rise a few thousand meters on each side of I-70. At Eagle, the Eagle River wanders slowly through the soft Eagle Valley Formation before joining the Colorado River at Dotsero and plunging into the White River Uplift at Glenwood Canyon.

Dotsero

Dotsero ("dot zero" on the railroad) has Colorado's most recent volcano. The "town" is at the confluence of the Colorado River and the Eagle River, at the west end of the Eagle Valley and just upstream from Glenwood Canyon. About 4000 years ago, a small cinder cone puffed clinkers and ash all over the area, and a small basaltic flow oozed across the site of the present I-70 and dammed the Eagle River. The flow crossed the ice-age gravels, so it is certain the eruption occurred after the last glaciation. After the water backed up enough to overflow the dam, it quickly eroded along the south side of the valley where the flow nudged against the soft Eagle Valley beds. Today the area has a gravel-mining operation, a cinder block factory and a campground, all contributing to the scenic confusion of the area, and causing most drivers to drive on, not noticing the rather interesting geology of the area. Don't get your hopes up to spot the majestic Dotsero Volcano. The cinder block factory has used up a lot of the cinder cone, and a small side road behind the trailer village of Dotsero leads up to the remains of old Dotsero Cone — a pile of cinders. Watch for the large trucks that are rapidly removing the remains.

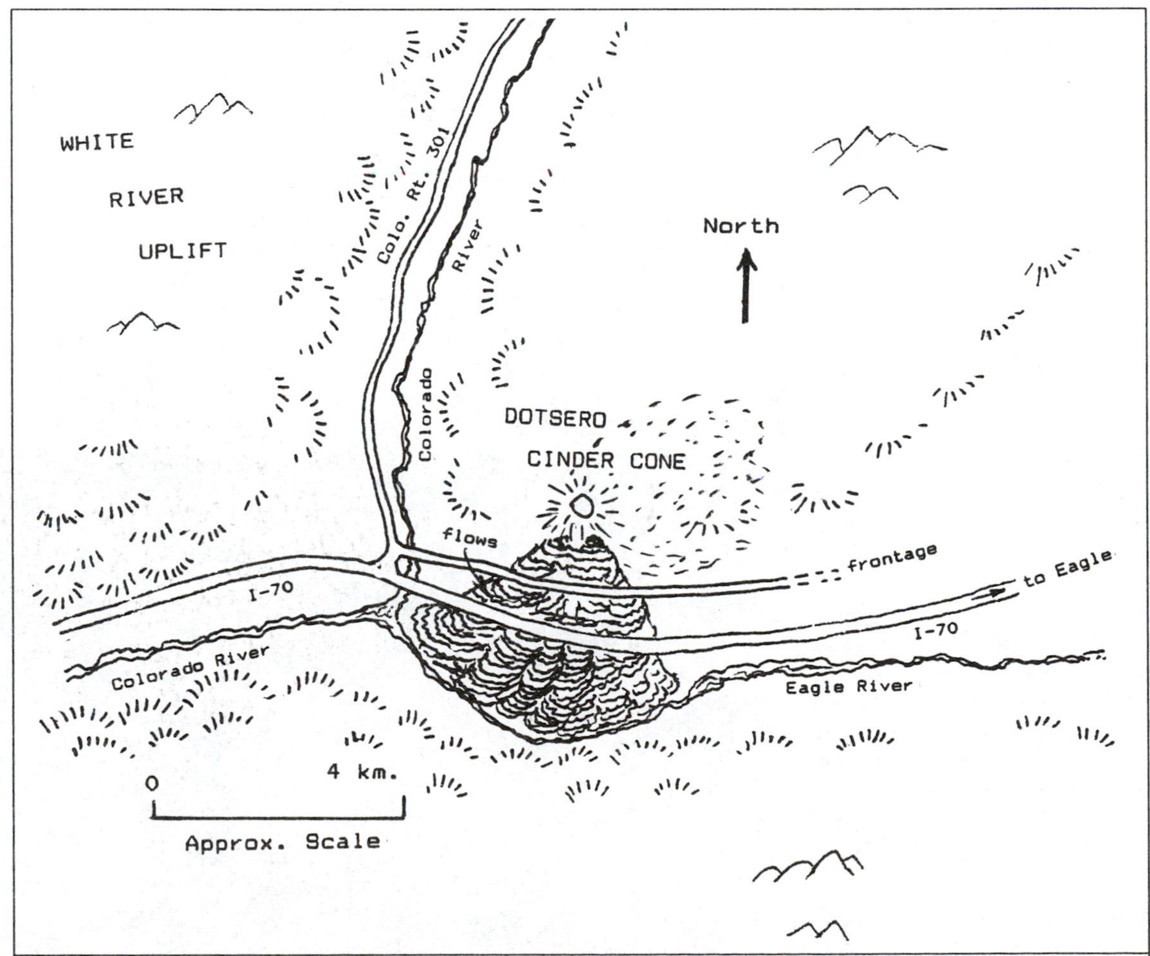

Fig. 155. Dotsero cinder cone spewed a pile of clinkers over the side of the Eagle River at the confluence of the Colorado and Eagle Rivers. Most of the cinders drifted east of the cone in the prevailing wind, and basaltic lavas flowed down the south side, damming the Eagle River about 4000 years ago.

Fig. 156. This is the clinkery toe of the Dotsero basalt flow. The Eagle River is near the cottonwood trees in the background, against the south side of the canyon. When the stream breached the lava dam, it incised its new channel against the soft gypsum-rich beds on the south wall.

The town of Gypsum is near the west end of the Eagle Valley, and gets its name, of course, from the gypsum beds that are exposed around the valley.

The folding of the gypsum layers suggests that the valley itself is an eroded salt dome. Soft, pliable salt, mixed with gypsum, squeezed upward in this crustal zone between the Gore Range and the White River Uplift, forcing the overlying beds into a steep anticline. The salt and gypsum are soft and easily eroded, thus allowing the Eagle River to carve a broad valley at the axis of the upfold. At Dotsero the rock layers on the south side of Eagle River tilt steeply to the south, and those on the north side of the valley are badly distorted, but dipping steeply to the north. Perhaps the structural weakness at this abrupt fold is responsible for the vent that fed the Dotsero cinder cone and associated flow. Some of the gypsum at Dotsero has been roasted and dehydrated where it is in contact with the cone.

The Eagle-Gypsum Products Company at the town of Gypsum was producing about 300 kt of gypsum per year in 1993, which makes about 1 million board feet of wallboard per day.

Fig. 157. The Dotsero cone is not a big feature. Here the clinkers in the background dwarf the vehicles in the lower right, but as volcanoes go, it is rather puny. White beds of the Eagle Valley Formation are pushed up to a nearly vertical attitude at this site just west of the trailer village of Dotsero.

Fig. 158. A group of geology students (circled) examines the roasted gypsum exposed below the cinders of Dotsero volcano.

Glenwood Canyon

Glenwood Canyon is a deep slash through the White River uplift, carved by the Colorado River. Cambrian, Ordovician, Devonian, Mississippian and Pennsylvanian sedimentary rocks are exposed in the canyon, in full view to motorists on I-70 and the main line of the Denver and Rio Grande Railroad. Planning for an interstate highway through the scenic gorge was delayed by many who felt that the canyon was too elegant to have an interstate highway to desecrate it. The resisters were a hundred years too late, however, because a highway, power lines, a railroad and power dam were already there when musician John Denver joined the resistance by symbolically throwing a silver dollar across the river in the canyon.

Construction of the Interstate took a decade, and the last link, the tunnel bypass at the Hanging Lake trailhead, was finished in 1992. For just a highway, the result is beautiful! Great pains were taken to preserve the scenery of the canyon. Many trees were marked which were to be saved during construction. Heavy equipment operators that might carelessly damage a marked tree would incur a heavy fine to the contractors. The alignment of the road was carefully planned to "fit" the topography, and "earth tone" concrete was used in the elevated spans. The task was monumental, and a frustration to tourists who often faced delays of over an hour while waiting for workers in the construction zones. The tourists were more than frustrated one day when a shower of boulders damaged several cars and killed one person.

Before the interstate, traffic would crawl through the canyon in a 30 mph (48 kph) speed zone. Many drivers challenged the slow limit and were fished out of the river in various physical conditions. Rock slides were common, then as now. The author remembers a wintry day when a train was blocked by a small avalanche that had tumbled down the shady side of the canyon. The train stopped safely, only to have a second avalanche block the tracks behind.

The Colorado River was already draining the area before the rise of the White River uplift, in the early Tertiary. As the uplift began, the river incised its channel a little at first, preventing the river from spilling out of the channel and picking an easier route to the west. The river had plenty of gradient and enough scouring material to keep ahead of the uplifting, and chiseled the deep gorge we see today. Some of the cutting was in Precambrian gneiss, similar to that in the Black Canyon of the Gunnison—some of the hardest rocks that a river can tackle.

The town of Glenwood Springs is at the west end of Glenwood

Canyon, where a group of faults cut the area. The earth's crust at this site has a high "GEOTHERMAL GRADIENT," meaning that a well drilled here can reach hotter rocks at a shallower depth than is normal. Water that percolates downward along the faults is heated, and rises convectively to the surface where it becomes the famous hot springs of the area. When the railroad was completed through the canyon, in the 1800's, many of the patrons of the rails came to Glenwood Springs for the "miraculous healing properties" of the hot mineral waters. According to a 1973 folder from the Hot Springs Lodge, the springs were developed in 1888, with the large pool 405 X 100 ft. (121 X 30 m.). The springs flow 3.5 million gallons (13.3 million liters) per day, at temperatures between 86 and 102 degrees Fahrenheit. Heavy mineral concentrations include sodium chloride, lime sulfate and potassium sulfate. Hydrogen sulfide gives the strong odor that is characteristic of many hot springs.

Glenwood Springs is actually a great number of springs, some of them developed, such as Vapor Caves and the big Hot Springs Lodge, but other springs bubble up in the middle of the river and others gurgle out of the river bank. Much of the geothermal potential of the area is currently wasted.

Each of the hard rock sequences makes a hogback on the west flank of the White River uplift. The late Cretaceous Mesaverde Group is approximately 1200 m. thick, and the layers are pitched up to almost vertical position. The Mesaverde is the most impressive of the hogbacks and is called the GRAND HOGBACK between Craig and the vicinity of I-70. Rifle Gap is a spectacular canyon where Rifle Creek cuts through the hogback. It is such an unusual canyon that the "artist" Cristo stretched a bright orange curtain across the narrows of the canyon in the 1970's. The wind and souvenir seekers promptly disposed of the curtain, but the environmental impact of the concrete anchor blocks remain.

Thick coal beds have been mined from the Mesaverde and its equivalents throughout Colorado. A series of mining disasters in the early 1900's killed about 80 miners, and a monument has been erected in their memory near the I-70 exit at Newcastle. The coal seams still burn slowly within the Grand Hogback at Newcastle, and the outcrops of some of the sandstones are so hot that snow cannot accumulate on them and the summer grasses near them are badly stunted.

The western margin of the White River uplift is marked by a splendid set of "page markers." The youngest rocks involved in the anticlinal fold of the White River uplift are the Wasatch Formation, of Paleocene-early Eocene age. Abutting the folded Wasatch, horizontally-bedded shales of the Green River Formation are exposed along Route 13, between Rifle and Meeker. The Green River is Eocene in age, and undisturbed, therefore the folding was rather short-lived and occurred just before 50 million years ago. The uplift disturbed the drainage of three states, resulting in the deposition of over 300 m. of lake sediments, including thick oil shales, in Colorado, Utah and Wyoming.

Fig. 159. Glenwood Canyon is a scenic, and famous corridor through the Rocky Mountains. There are relatively few places where a railroad could have been built, reasonably, in the 1800's, and this canyon was one of them. This pre-1980 photo has no interstate highway with its "earth-toned" state-of-the-art elevated sections. Some tourists after 1992 are more impressed with the highway than the geology, which is still there in its full splendor. The Cambrian Sawatch standstones are most of the wall shown here. The top vertical ledge is Ordovician Manitou Formation.

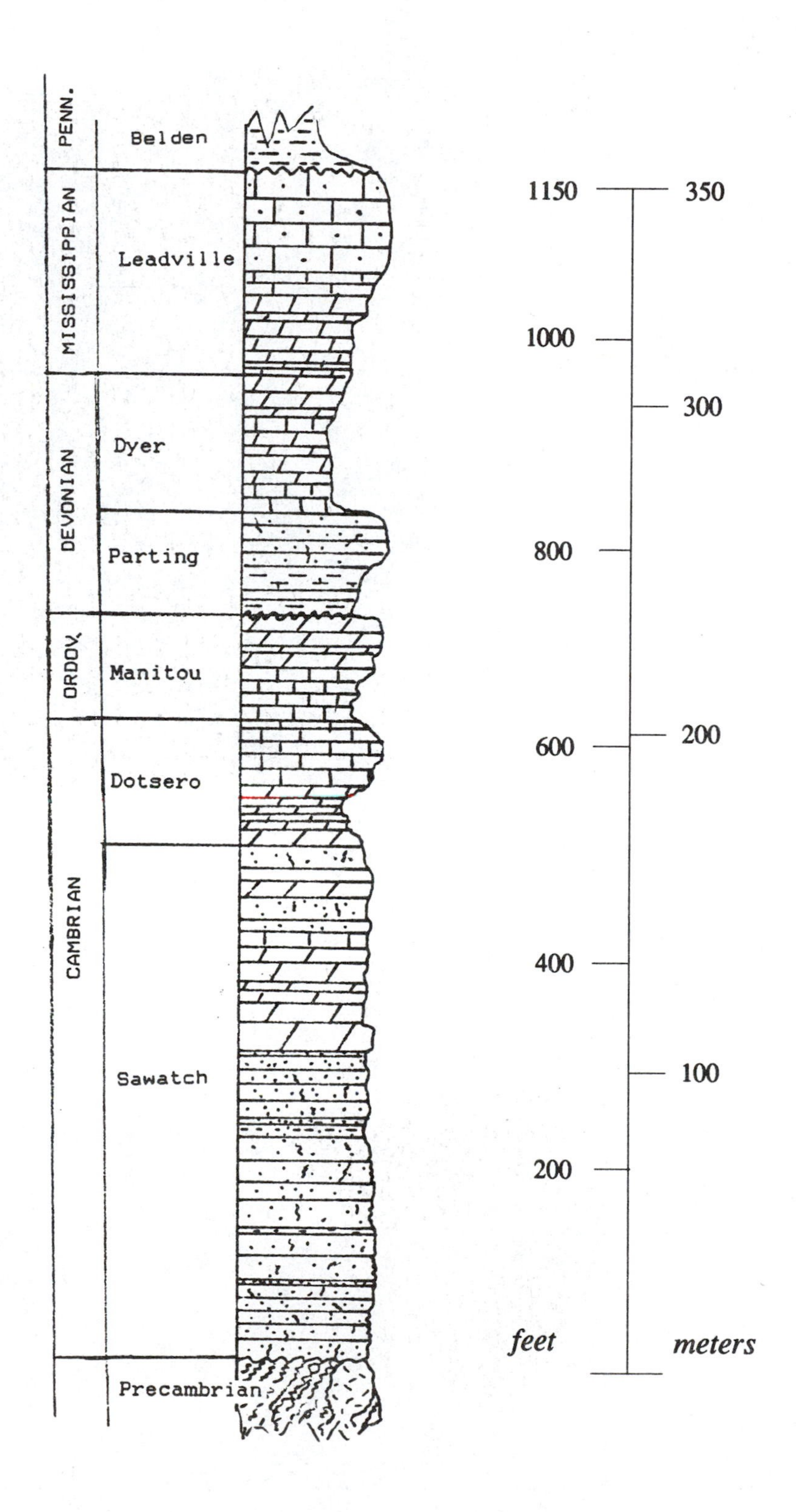

Fig. 160. A simplified stratigraphic section of the rock sequence at Glenwood Canyon, plus the younger units exposed in the nearby areas.

Fig. 161. This winter scene in Glenwood Canyon has more than 300 meters of Precambrian metamorphic rocks exposed in the bottom half of the photo. The site is approximately the axis of the White River uplift.

Fig. 162. A freight train of the Rio Grande Railroad has been stalled by a small avalanche that has run down a notch in the Precambrian section near the Hanging Lake trailhead.

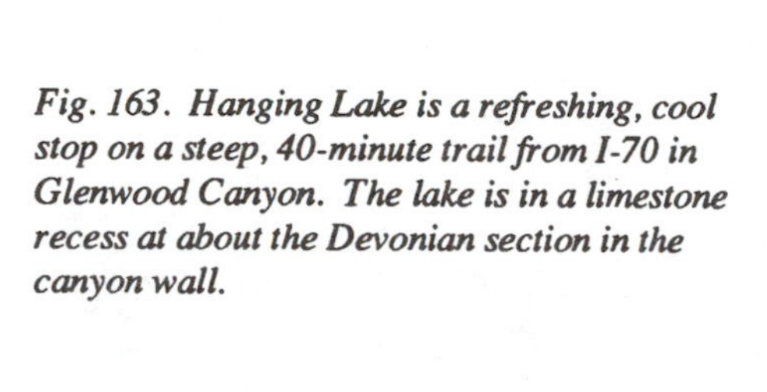

Fig. 163. Hanging Lake is a refreshing, cool stop on a steep, 40-minute trail from I-70 in Glenwood Canyon. The lake is in a limestone recess at about the Devonian section in the canyon wall.

Fig. 164. Streams that feed Hanging Lake include these two spouting holes.

Fig. 165. Limestones of the Mississippian Leadville are pushed up at the west end of the White River uplift as seen in this view from West Glenwood. Soft shales of the Belden Formation overlie the Leadville, so the Leadville is typically the top of a series of cliffs that extend all the way down to the Cambrian Sawatch Formation, a section of about 400 meters.

Fig. 166. Mineralized water at 100 degrees Fahrenheit supplies the huge swimming pool at Glenwood Springs, providing a favorite recreation site for over 100 years.

Fig. 167. The Colorado River during low water stage in September reveals travertine cones in the middle of the river, where hot, mineralized water feeds the stream. The cone in the foreground is on the south bank of the river.

Fig. 168. Close-up view of a travertine cone in the middle of the Colorado River at Glenwood Springs. The flow from this cone was only about five liters a minute in 1975.

Fig. 169. Near the west entrance to Glenwood Canyon, hot, smelly mineral water bubbles through green and yellow algae and into the Colorado River alongside the railroad. Old wooden and tile pipes in the mess suggest an early attempt to exploit this copious flow.

Fig. 170. Eastbound travelers on I-70 see this exposure of the Maroon-to-Dakota section near the town of Silt.

Fig. 171. Westbound travelers see this colorful view near Silt. The Maroon is a deep red, with a white sandstone sequence at the top, followed by more bright reds of the Chinle. A thin, pink cliff is called the Entrada here, overlain by the multi-colored Morrison Formation. Hard, brown, siliceous Dakota forms the resistant ridge (hogback) on top.

Fig 172. Between West Glenwood and Newcastle is this exposure of red Maroon Formation beds which are tilted as the west flank of the White River uplift -- or the east flank of the Piceance Basin.

Fig. 173. Thirty kilometers south of Glenwood Springs is Mt. Sopris, at 3885 m., a small stock of Tertiary intrusive rock that has been sculptured with intensive glaciation. The view is from Highway 82 at Carbondale, along the Crystal River.

Fig. 174. East of Newcastle the uplifted Mancos dips away from the White River uplift at about 35 degrees. In a textbook example, horizontal gravel beds of Pleistocene age cover the truncated top in an angular unconformity.

Fig. 175. A small underground mine near Craig was removing a 5-meter seam of coal in the 1970's. Some large strip mines operate in the Craig area to support coal-fired steam generators at Craig and Hayden. The Hayden plant produces about 450 MW of power.

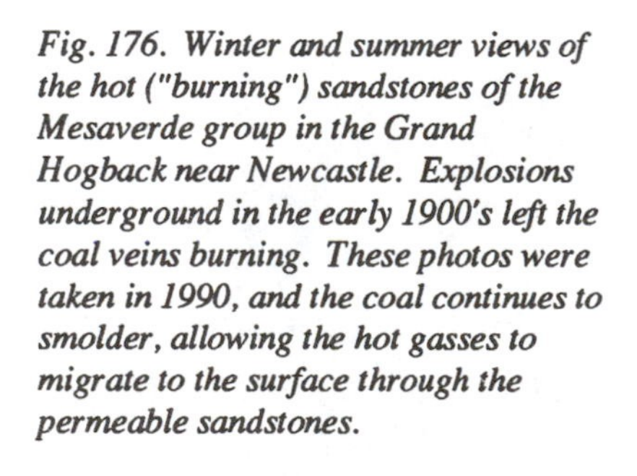

Fig. 176. Winter and summer views of the hot ("burning") sandstones of the Mesaverde group in the Grand Hogback near Newcastle. Explosions underground in the early 1900's left the coal veins burning. These photos were taken in 1990, and the coal continues to smolder, allowing the hot gasses to migrate to the surface through the permeable sandstones.

Fig. 177. Rifle Falls, 16 km. north of Rifle, plunges over a 10 meter travertine ledge. Occasionally, nice trout are found, stunned, at the base of the falls, having escaped from the hatchery just upstream of the falls.

Fig. 178. A person (white shirt) sits at the base of the network of travertine ledges at Rifle Falls. The falls have shifted position frequently in recent geologic time, building thick deposits of travertine at each location. The water of Rifle Creek contains much calcium carbonate, which precipitates when some of the water is evaporated at the falls.

Fig. 179. This exposure occurs about 3 km. downstream from Rifle Falls. The pink cliff is Entrada (?) and apparently ground water was able to migrate laterally along the base of the sand, in contact with the impermeable red Chinle Shale below the sand. Organic material along the contact precipitated uranium and vanadium minerals in the base of the sand. The circle marks a mine entrance blocked with a steel door. The vanadium and uranium were extracted at a plant in Rifle, leaving a big pile of radioactive waste to be cleaned up in the 1990's.

Fig. 180. The Dakota forms "flatirons" at the west margin of the White River uplift, as veiwed from I-70 about 5 km. east of Newcastle.

Fig. 181. An impressive Tertiary erosional terrace near Rifle.

The Piceance Basin

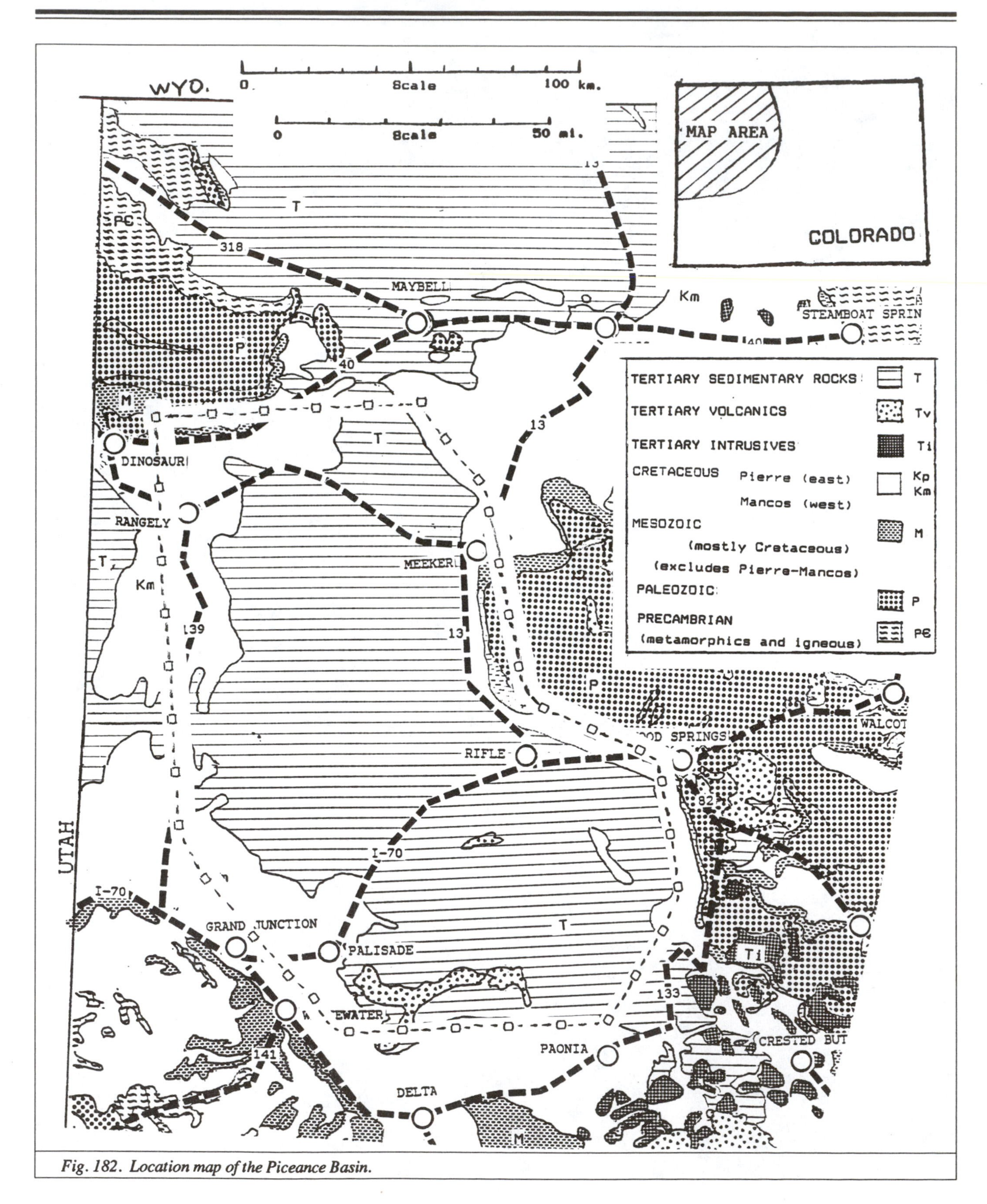

Fig. 182. Location map of the Piceance Basin.

The Piceance Basin is an asymmetrical syncline between the Grand Hogback-White River uplift on the east and the Uncompahgre uplift and Douglas Arch on the west. The basin plunges to the north and is about 130 km. long in a north-south direction and 80 km. east-west. The age in the basin is rather precisely marked along the Grand Hogback. As early as the King survey of the 40th Parallel (1877), the steeply folded Wasatch beds of the Paleocene age were recognized adjacent to undisturbed beds of the Green River Formation of Eocene age. Uplift of the White River Plateau apparently helped to enclose the basin (Piceance) that became the depositional site of the Green River shales. In fact, three basins developed. The Piceance, the Uinta Basin, mostly in Utah, and the Green River-Washakie Basin of Wyoming. The Green River Formation is named from outcrops at Green River, Wyoming.

Good bituminous coal underlies the Piceance Basin in Colorado. Mines in the Mesaverde Group along the Bookcliffs and Grand Hogback have penetrated several miles under the basin and have supported the economy of western Colorado since the first mines began operations about 1889. The transcontinental railroad first linked California to the eastern U.S. in 1869. The early trains provided a market for the coal, as fuel for the steam engines, and as a means of delivery to industrial sites. Diesel engines replaced the coal-fired steam engines in the 1950's, but the worldwide appetite for good coal guarantees a continued market for Piceance coal—even though the demand is extremely cyclical. Booming U.S. or world economies require peak production and most mines produce at full capacity, but recessions will close most of Colorado's mines. The Cretaceous Mesaverde coal is very low in sulfur, typically less than 0.5%. This compares favorably with Pennsylvanian coals of the eastern U.S. which range from 4% to over 15% sulfur. Mesaverde coal seams are relatively thick also. Two-meter thick seams are common and are easily mined, especially where the beds are horizontal or only gently tilted. Beds over 10 m. thick are mined in several districts.

Natural Gas is also contained in the Mesaverde throughout much of the Piceance Basin. The reservoir sands are relatively "tight" and do not easily release the gas to production holes. Two remarkable pioneering tests were made on gas-bearing sandstones of the Mesaverde in the Piceance Basin. The first, in 1969, was Project Rulison, where a 43 KT (equivalent to 43 thousand tons of conventional TNT explosives) nuclear device was detonated near Rulison at a depth of 2528 m. The test was an engineering success with good stimulation from a huge rubble "chimney" but the gas that was tapped came up disturbingly radioactive with tritium. A second much more ambitious test, near the center of the basin, used three 30 KT devices at depths of 1752, 1869 and 2007 meters. Again the test was getting perhaps 10 times the stimulation of conventional methods, but the gas in the chimney was radioactive. Also some buildings were damaged 48 km. away in the Rangely area. A chimney toppled from the museum at Montrose, a full 128 km. away.

Some mining people are concerned with shaking up and possibly fracturing the overlying Green River Formation which has a greater economic potential than the Mesaverde gas. Oil shale in the Piceance Basin eclipses any other commodity in the state for its potential value. The 1990 crude oil reserve for the United States was about 25 billion barrels. Those are

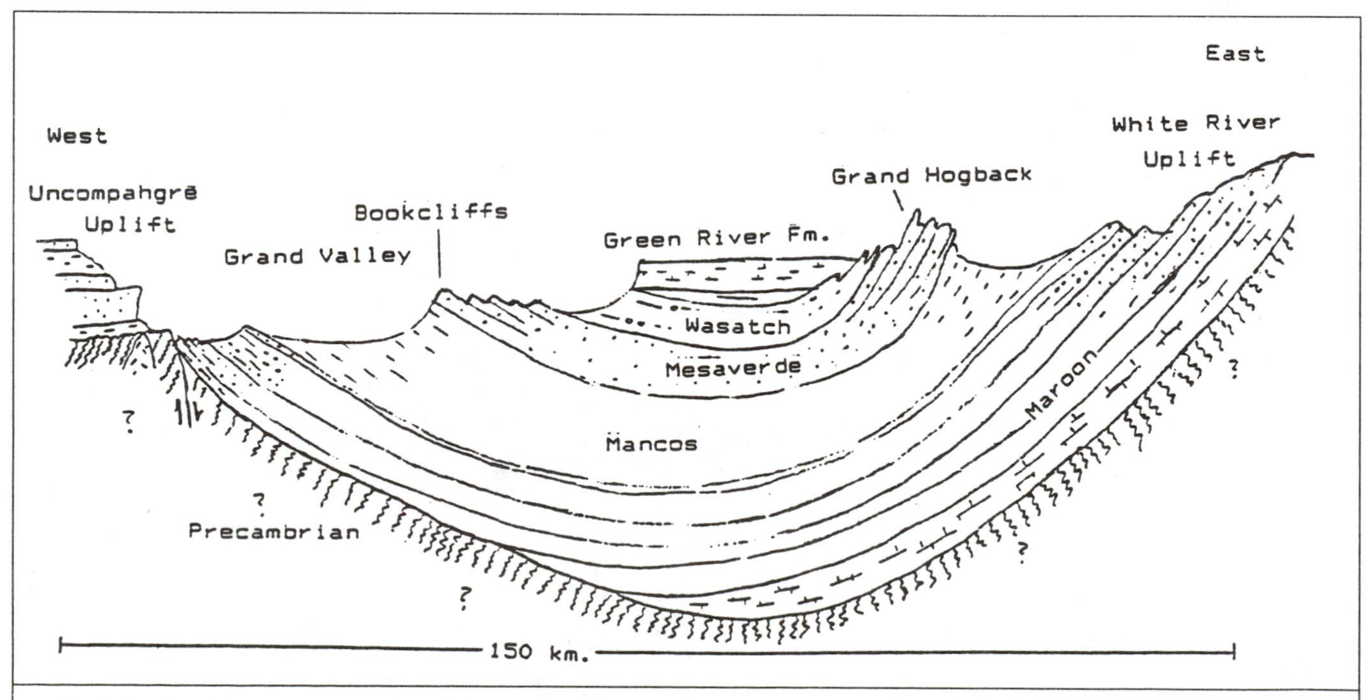

Fig. 183. Simplified crossection of the Piceance Basin from the Grand Junction area on the west to Glenwood Springs on the east. Note that the Maroon Formation pinches out toward the west.

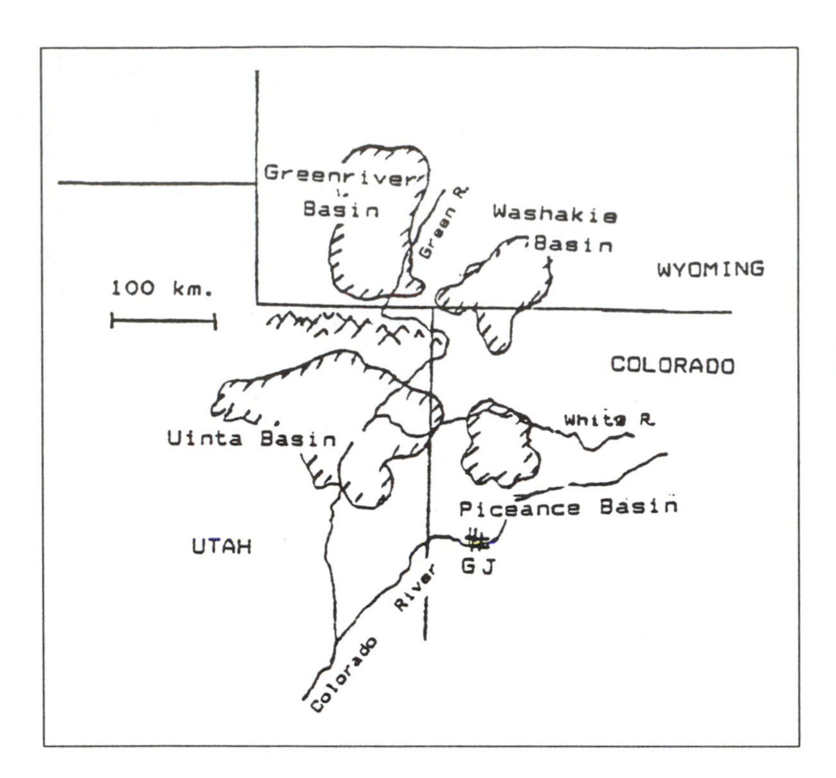

Fig. 184. Map of Eocene Lake basins that contain significant oil shale.

proved reserves in all states, including familiar oil states such as Texas, Alaska, California, Oklahoma, Louisiana and counts the offshore areas. The oil shale deposits of Colorado, Utah and Wyoming are in the trillions of barrels. Trillions of barrels of oil in the deposits is actual fact, but not very realistic. There is lean oil shale and there is very rich oil shale. If the shale is roasted (retorted) and produces less than 20 gallons of oil per ton of rock it is not rich enough to interest the industry. Therefore, the many trillions of barrels that may be reported are not very realistic numbers. However, the rich "mahogany zone" of the upper Green River Formation has many square miles with oil shale that can produce better than 25 gallons per ton as an average grade from more than 30 m. of vertical section. Colorado alone has over 600 billion barrels of recoverable oil in these very rich beds. Shale has been mined from several leases using huge equipment that "eats" a 9 m. high room—then comes back for a second pass to take another 9 m. out of the floor. Several kinds of retorts have been tested that can cook the raw shale to the 900 F. needed to drive out the oil.

Why does the oil shale giant lie dormant? Cost! Union Oil's retort at Parachute produced up to 7000 barrels per day until 1991, but it had a guaranteed price contract with the government and still went over budget. The 1973-80 boom in the oil shale industry was caused by foreign crude oil prices that climbed rapidly to over $30.60 per barrel. At $35-40 per barrel the extra costs of mining, retorting and disposal of the spent shale can be paid and leave some profit for the producer. In 1988, the price of crude oil slipped on the world market to below $20 per barrel and snuffed out the oil shale spark. The same

Fig. 185. An experimental oil shale mine operated between 1955 and 1982 at this site at Anvil Points. More than 35 meters of shale containing an average over 25 gallons per ton of rock can be mined at this location. The road accommodated 6 meter wide trucks, and the portals are 10 meters high. Inside the mine, the cavernous operations are over 20 meters high. Cost of producing oil this way is over $34 per barrel.

Fig. 186. A geology class visits the U.S. oil shale experimental retort operations at Anvil Points. The facility was removed in the 1980's and the land restored to former conditions.

Fig. 187. A 6-meter wide truck arrives at the portal of the Anvil Points operation.

boom-bust cycle has been repeated many times in the past. However, the U.S. conventional oil reserve continues to decline, and if cheap foreign oil gets too scarce or too expensive, the oil shale industry will have to be revived.

Oil shale is actually a misnomer. It is not shale and it does not contain oil. The unique Eocene basins of Colorado, Wyoming and Utah were filled with huge freshwater lakes. Vegetation in and around the lakes was luxuriant, and a rich diversified biota of fish, insects, plants, etc. abounded. But the deeper portions of the lakes had poor circulation, and the material that sank to the bottom was not scavenged effectively and decay was retarded or halted completely. Instead of the clean sand, clay, and limestone that fill most sedimentary basins, these unique Eocene lakes deposited carbonates—limestone, and especially, dolomite $CaMg(CO_3)_2$ mixed with clay and vast quantities of organic matter. As organic marlstone, the thousands of meters of sediment that accumulated there had little compaction because carbonates make a rigid framework. The organic material became entombed with little change, and in time was only partly altered in the chain of events that is needed to produce crude oil. "Kerogen" is the name given to the waxy, immature "oil." Most oil source beds are compressed over time and the oil ingredients migrate out and maturate to become crude oil.

So, the stuff is not oil shale, but "kerogen-rich marlstone." Near-perfect fossil specimens of leaves, fish and insects have been recovered from a number of Green River collecting sites. A world class site at Kemmerer, Wyoming has been called the richest fossil fish location in the world.

Conditions in the lakes reached their zenith at the time of the Mahogany Zone of sedimentation. Annual varves (paired layers) formed. The black kerogen was deposited each year,

Fig. 188. This view of Anvil Points is from the mine portal. The hard cliff is the mahogany zone, which is the richest oil shale. Over 70 gallons per ton has been calculated from the richest shale. An average of 25 gpt is needed for "commerical" grade shale.

Fig. 189. Exxon's operation up Parachute Creek from the town of Parachute, is massive. The roadway shown here is a 4-lane access for the huge haul trucks. This large gash is not the mine, only the road to the top of the mountain where the retorts are to be sited. Retorts on top will have a better air circulation than would be expected for retorts in the valley, where temperature invervsions can stagnate the air in winter.

covered by a light-colored, carbonate-rich layer. These distinctive laminae are less than 2 cm. per pair and may contain up to 70 gallons of oil per ton of rock in the richest zones. The environment lasted thousands of years. To make the whole system easier to evaluate, a nearby volcanic event spread an ash layer over all three states, leaving a distinctive, hard layer called the mahogany marker. Miners use the marker as a handy referece for the richest shale section.

As the oil shale environment eventually waned, evaporation in the basin exceeded the infill and the sedimentation culminated in hundreds of meters of salt and other evaporite minerals. Remarkable deposits of salt, nahcolite ($NaHCO_3$) and dawsonite ($NaAl(CO_3)(OH)_2$occur near the deepest part of the basin, reaching 300 meters thick as a maximum. Nahcolite is known as sodium bicarbonate to the layman.

Fig. 190. The Colorado River and I-70 cut deeply into the south end of the Piceance Basin, exposing Green River oil shale and the underlying reddish shales, mudstones and gravels of the Wasatch Formation. This view is from I-70 between DeBeque and Parachute.

The Grand Mesa

Fig. 191. Columnar joints in volcanic rock.

During Miocene time, about 9.5 million years ago, extensive flood basalts oozed from fissures in the location of the present Grand Mesa. The fluid basalts filled the adjacent stream valleys up to 200 m. thick of repeated layers of lava, oozing a few feet here, then spreading to another low spot, returning to the first area later, and continuing for many days, weeks, and possibly centuries. Eroded surfaces and soils separate some of the individual flows, and some of them have the distinctive columnar jointing that typifies extrusive rocks. Although the extrusive rock is classified as "volcanic," most geologists would not call the Grand Mesa a volcano. Fissure flows normally make flat terrain, such as the Columbia River basalts of Washington State or the Snake River plains of Idaho. The basalts on the Grand Mesa include vesicular textures, scoria and dense rock that is nearly bubble-free. Because of variations in the amount of oxidized iron, the rock may be black, brown, red or gray. After the basalt flows filled the valleys, the entire Colorado Plateau (western Colorado and its neighboring states), was slowly uplifted in a vast bulge, rising more than 2,000 m. above the former surface. The basalt that filled the valleys became the most resistant rocks in the vicinity, and quickly became isolated as nine million years of relentless erosion removed the surrounding softer materials. The present outline of the Grand Mesa occupies about 2600 square kilometers, or about one

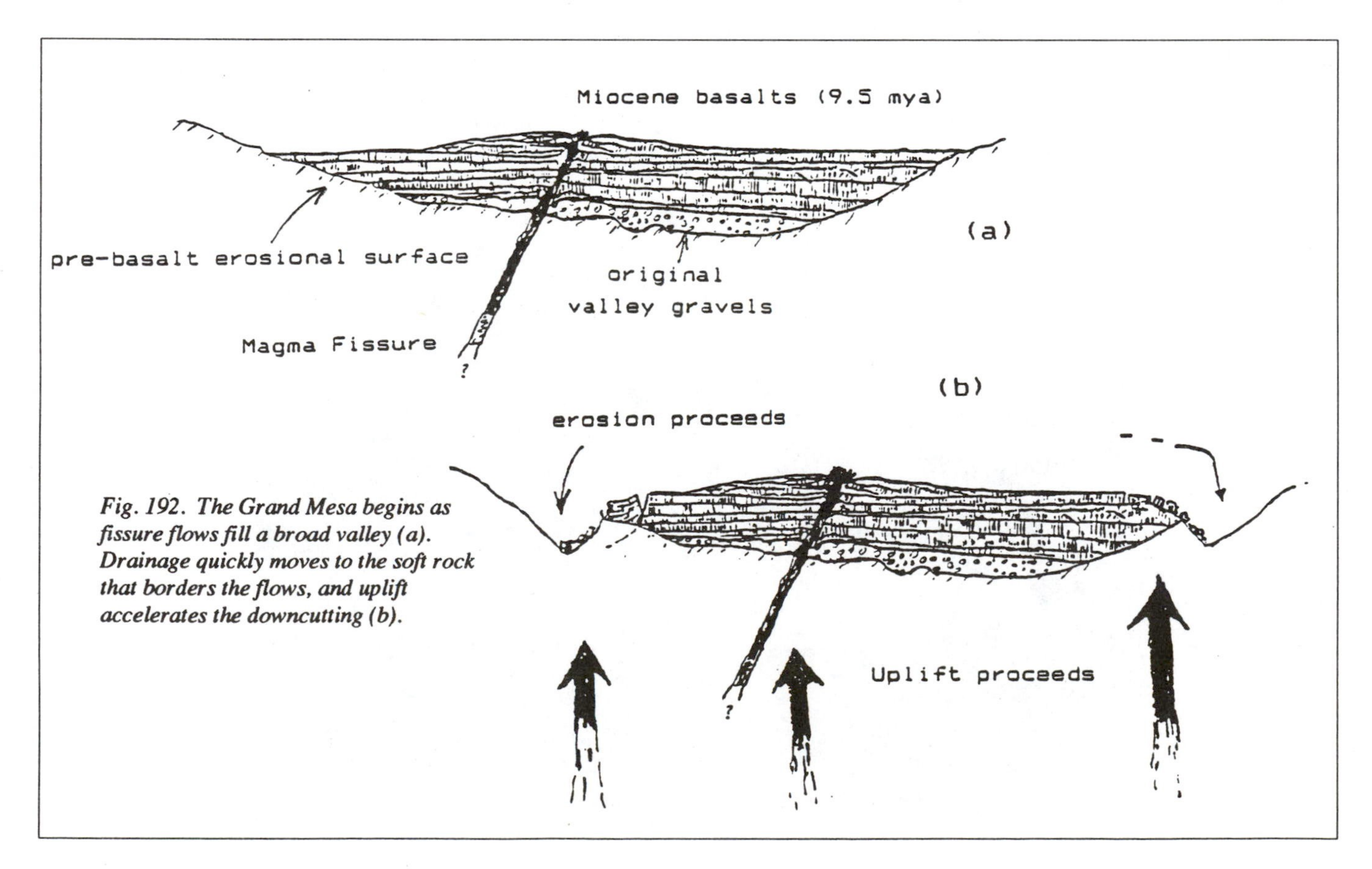

Fig. 192. The Grand Mesa begins as fissure flows fill a broad valley (a). Drainage quickly moves to the soft rock that borders the flows, and uplift accelerates the downcutting (b).

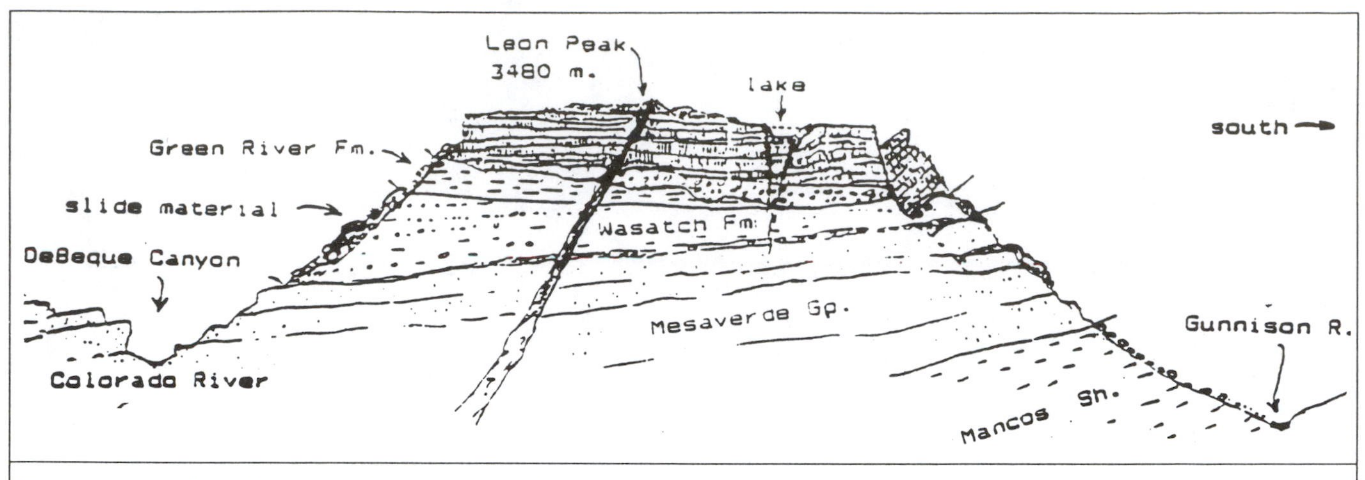

Fig. 193. After nine million years of erosion, a lava-filled lowland becomes a high plateau (oops, a MESA!).

percent of the area of the entire state. Some of the local tourist information refers to the Grand Mesa as the largest flat topped mountain in the world—or the largest mesa in the world. This is really a bit misleading because surely Tibet and Greenland are larger, flat-topped features, and if you have the world's largest mesa, perhaps it is the world's smallest plateau.

The basalt flows that cap the mesa have vertical joints that are typical of extrusive rocks. These joints allow water to percolate through the basalts and reach the underlying shales of the Green River Formation. There is a tendency for the water to flow laterally along the top of the underlying clays.

Some of the mesa is also underlain by a system of stream gravels. The gravel allows ground water to move toward the exposed edges of the lava flows, and springs are common in, and under, the basalt cliffs around the mesa. The water also softens the clay, and the saturated clays tend to slump and slide down the sides of the mesa. As the underlying clay slides away, the basalt columns are left unsupported and thunderous rock slides can occur as the vertically jointed lavas tumble down the steep slopes. In some cases hundreds of meters of the mesa's edge will slip away, at speeds that may be imperceptibly slow, or range upward to rapid and violent. The thick and brittle basalts, resting on rather squishy, moist clay shales causes parts of the mesa to develop deep cracks that can open and form large troughs. If the underlying basalt is not too porous, lakes can fill the basins, and there are dozens of premier trout fishing spots on the mesa. A large elk herd lives there along with plenty of deer and other wildlife.

Slopes of the Grand Mesa are usually "HUMMOCKY" (with many lumps and mounds) due to the slope failures of various types. The north-facing slopes are the most slide-prone because the shady sides of mountains retain moisture better than the sunny sides. The moisture helps to lubricate the clays

Fig. 194. Lands End Road descends the west end of the Grand Mesa.

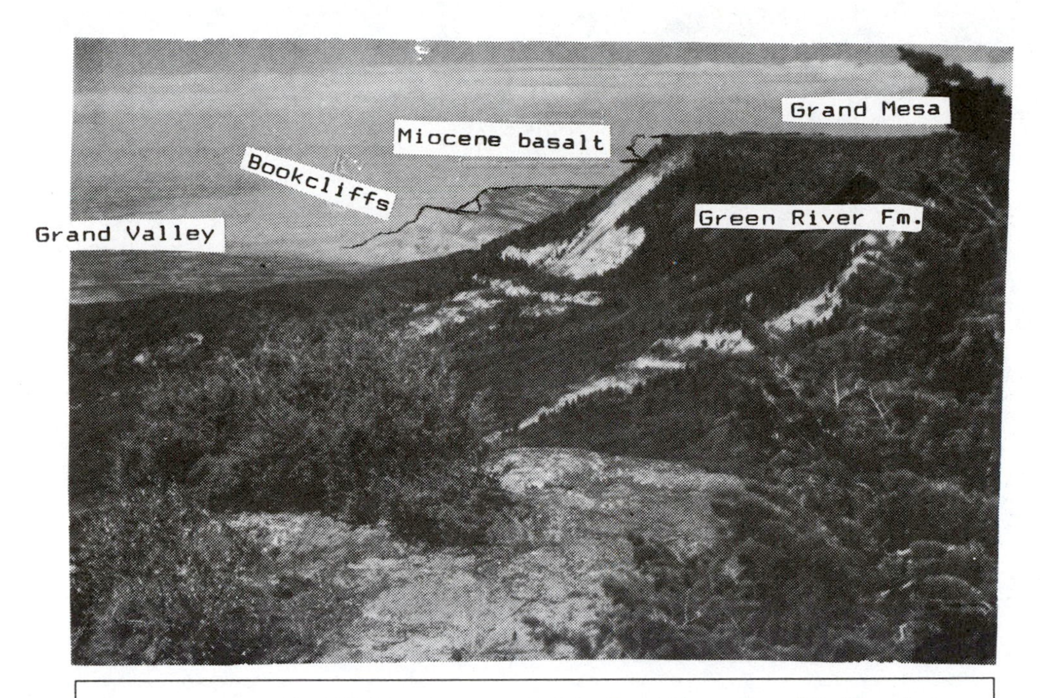

Fig. 195. View looking NW from the west end of the Grand Mesa. Top of the Mesa is about 3070 m.; top of the Bookliffs is about 2150m. Level of the Grand Valley is about 1415m. Debris slides have left white scars that expose the Green River Shale which underlies the basalt flows that cap the mesa.

Fig. 196. Wolverine Lake nestles in a trough opened by fractures in the basalts of the Grand Mesa.

Fig. 197. Grand Mesa basalts are about 9.5 million years old and range from scoria to very dense. Variations in weathering and hematite give colors from black to dark, grayish-red. Entrapped gas bubbles cause this typical vesicular basalt.

and also adds more weight to the unstable material. Those parts of Colorado that are more than 2700 m. (9000 ft.) above sea level were glaciated. The Grand Mesa had a large ice cap during the Pleistocene, and some of the ice scraped over the edges and slid down the shady sides. Landsliding has obscured most of the features that the glaciers would have left, but at the town of Mesa, on the north side, there are some suspicious features that are probably outwash material and may include some material left directly by ice glaciers. Glacial scour has been found on basalts on top of the mesa.

Fig. 198. These boulders of vesicular basalt, in transit down the slope of the Grand Mesa, get a caliche crust on their undersides in the arid climate. These were turned over when Lands End Road was made.

Fig. 199. Basalts that cap the Grand Mesa rest on soft shales of the Green River Formation. The rim of the mesa continually recedes as the shale slope fails and creeps downward. Fresh scars are common, and unstable slopes are the rule on all sides of the mesa. Two old slides are outlined on this photo.

Fig. 200. Island Lake on the Grand Mesa fills a depression formed by adjustments in the lava pile as a result of the unstable rocks under the basalt. The dozens of well-stocked lakes on the mesa have good trout fishing, both for summer and (brrr!) winter ice fishing. Some of the lakes are "bottomless," which means the lake is deeper than the measuring tape. Lakes would be sealed by the squishy Green River Formation if they were deeper than the cracks in the basalt.

The Grand Valley

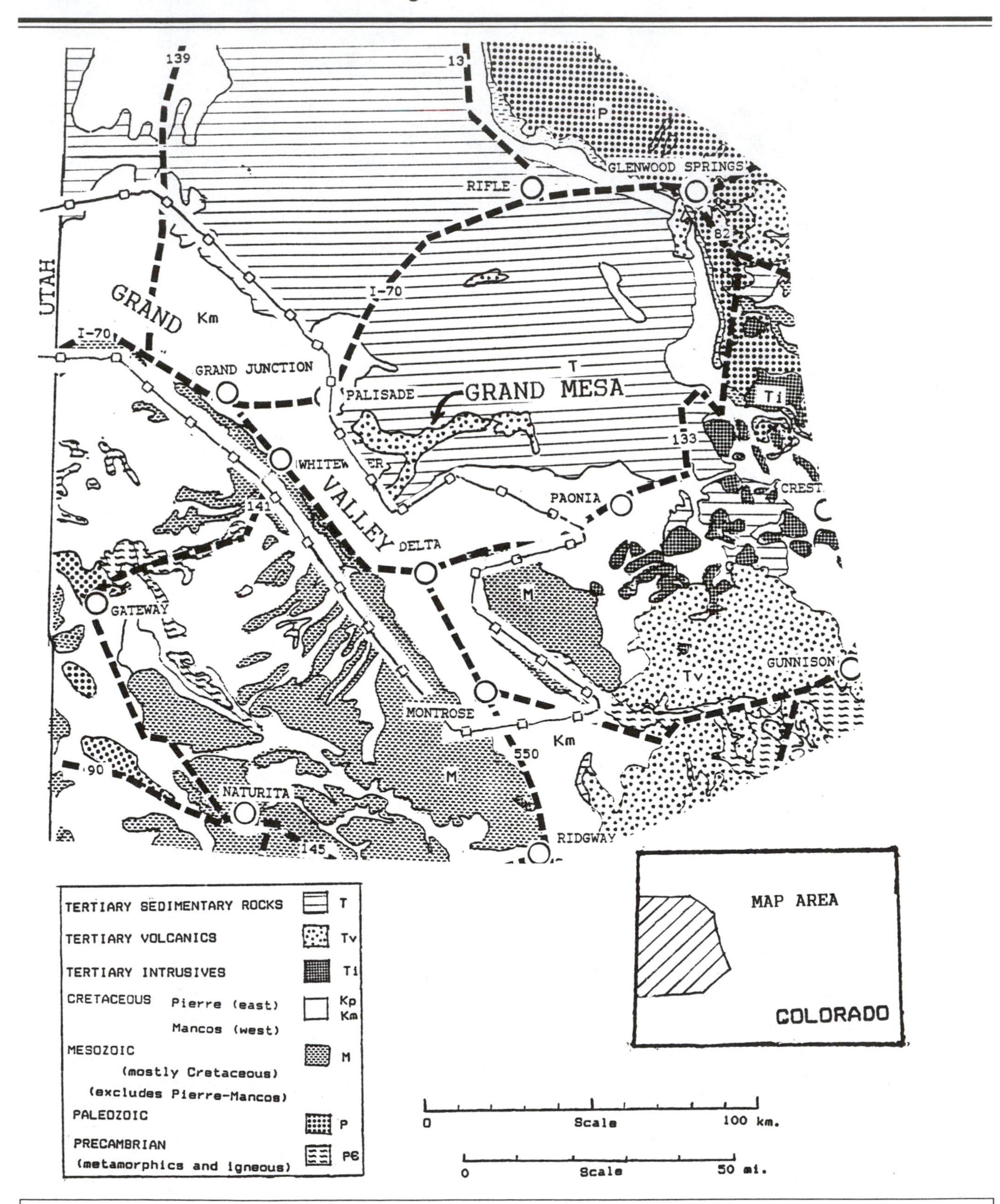

Fig. 201. Map showing the location of the Grand Valley and Grand Mesa.

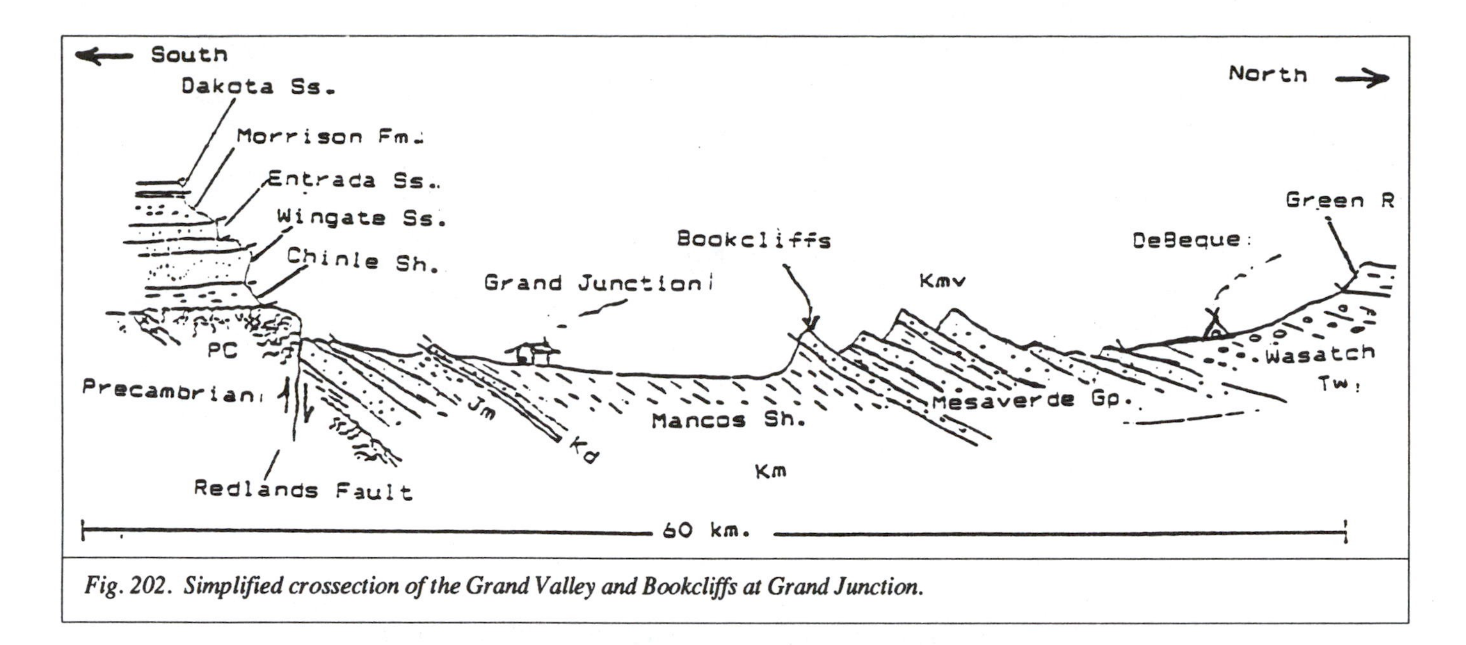

Fig. 202. Simplified crossection of the Grand Valley and Bookcliffs at Grand Junction.

The Grand Valley is a broad, open area between the Uncompahgre uplift and the Bookcliffs. With no real boundary, the "valley" extends to the south 100 km. through Delta, to Montrose. Westward it reaches 250 km. to Price, Utah, and swings south again to the deserts near Huntington, Utah. Total area of this Grand Valley-Green River desert is more than 13,000 square kilometers, but less than 10% of it is in Colorado, or about 1.2% of the state's area. The Grand Valley in Colorado gets about 20 cm. (8 in.) of precipitation per year, so many Coloradans are glad Utah got most of it. Tourists have stated that the Green River desert is pretty bleak!

Most of the Grand Valley is eroded from gently tilted Mancos Shale. The Mancos is a sticky (when wet) clayey mudstone which was deposited in a shallow marine environment that had much suspended material in the water. The Mancos, and its equivalents on the eastern slope (Pierre, Benton, Niobrara, etc.), are the remains of a seaway that covered all of Colorado and most of the adjacent states during late Cretaceous time. At Grand Junction the formation is about 1,200 m. thick, yet the bottom-dwelling microfossils that have been identified from the shales suggest that the water was never deeper than about 200 m. How can 1,200 m. of sediment be deposited in an ocean only 200 m. deep? That is a good question, but the answer is simple. As the sediment accumulated on the seafloor, the conditions that made the seaway in the first place continued to deepen the sea. The situation was maintained until there was enough sediment in the area to squeeze down to about 1,200 m. of hard shale.

Before the Mancos seaway invaded the area, the scene was a broad, continental environment, with meandering streams, ponds and a relatively dry surrounding terrain. The Morrison Formation, with all its dinosaur fossils, represents a system of streams and wide and shallow basins that frequently received dust from volcanoes erupting somewhere to the west. The area

Fig. 203. Riggs Hill, in Grand Junction is the site of the first Brachyosaurus ever found, in 1900. The nearly complete skeleton was found near the top of the Jurassic Morrison Formation.

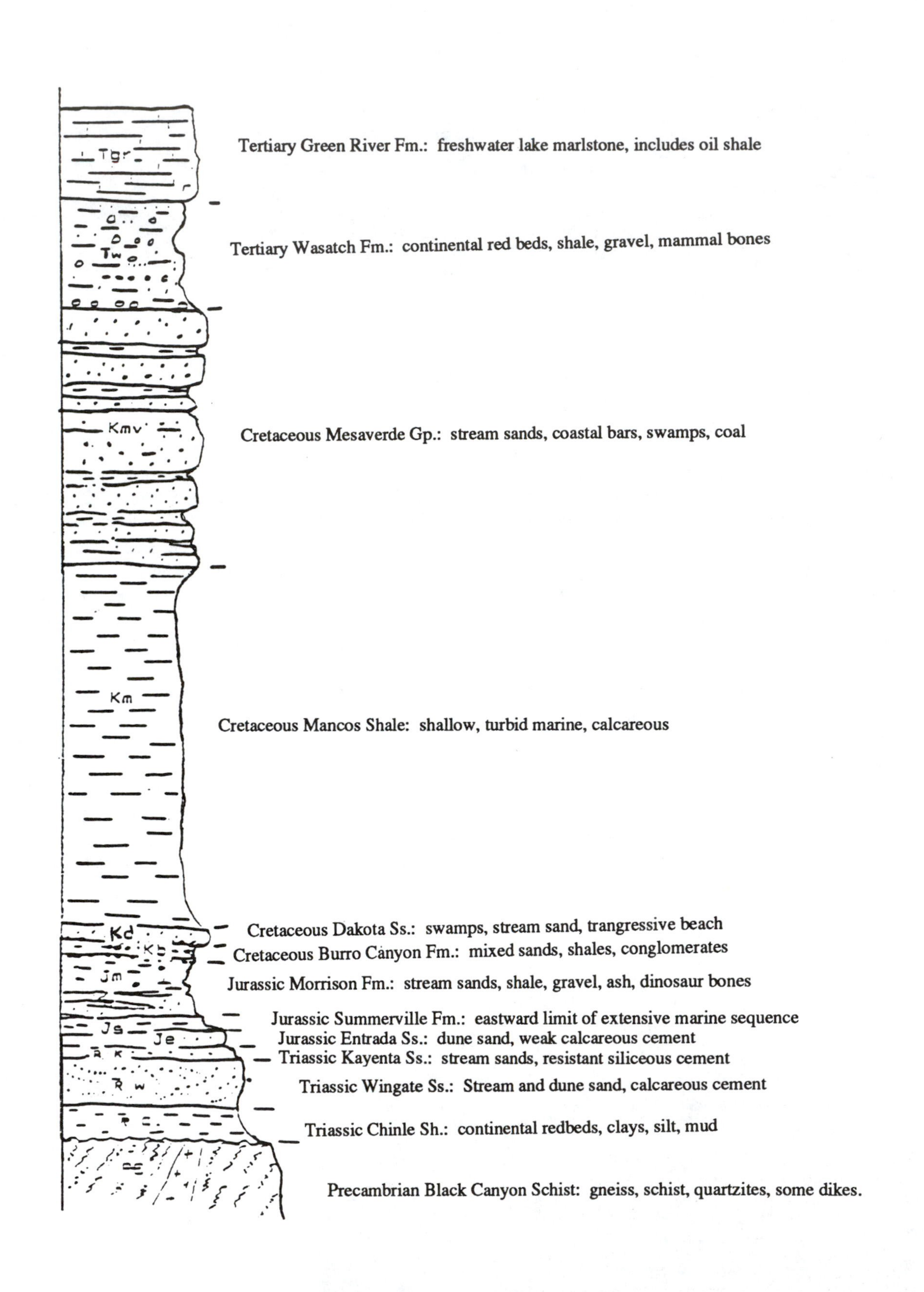

Fig. 204. Stratigraphic section at Grand Junction in the Grand Valley.

Fig. 205. A bronze plaque marks the site of the Brachyosaurus discovery.

hosted an amazing variety of plants and animals, including the dinosaurs, that have captured the imagination of all the world. When the Mancos seaway advanced over the land, the waves smashed the rocks of the shore into sandy beaches that are seen now as the Dakota Sandstone. Figure 203 is the stratigraphic section at Grand Junction. Some of the formations in this sequence occur statewide, with only a little modification in the names at various sites across the state.

In and near Grand Junction a number of important dinosaur bones have been recovered. In 1900 paleontologists from the Chicago Field Museum, including Elmer Riggs, discovered the world's first *Brachyosaurus altithorax* at Riggs Hill, which is within the boundary of Grand Junction. When it was discovered, it was the largest dinosaur known. Chicken-sized reptiles have been collected at the Fruita Paleontological Area, which is only 19 km. west of Grand Junction, and another 24 km. west, just off Interstate 70, at the Rabbit Valley exit, several dinosaur bones are exposed along a nature trail that is being developed. Dinosaur experts have included Grand Junction in the "Dinosaur Triangle" which links three important sites of dinosaur material. Dinosaur National Monument, 200 km. to the north and Cleveland-Lloyd Quarry, 240 km. to the west complete the triangle. Most of the dinosaur collections are from the Morrison Formation of Jurassic age.

Fossils are locally abundant also in the Mancos Shale (Km.). *Baculities* is found throughout the formation, especially in some concretions near the top of the unit at the base of the Bookcliffs. *Scaphites, Inoceramus, Prionocyclus* and the diminutive oyster, *Ostrea lugubris* are found in the Mancos. *Pychnodonte* is abundant in a key marker bed near the base of the Mancos and large ammonites, shark teeth and other shallow marine fauna are known from the formation. The Bird Bath locality, with its fine assortment of *Placenticeras* (p.77-79) is included in the Mancos interval. Grand Junction gets its name from the confluence of the Colorado and Gunnison Rivers. Before 1930, the Colorado River was officially the "Grand River," and the Grand Valley, Grand Mesa and Grand County

Fig. 206. This view of the Colorado River in 1980 shows what happens when the river reclaims its floodplain during a flood. With a good winter snowpack in the mountains, followed by a rapid spring warmup, the normal channels of the river are not enough to carry the load. The view is to the west from near the 32 Road bridge. The Gunnison River enters the stream from the left, near the top of the photo.

Fig. 207. Mount Garfield is near the east end of the Bookcliffs at the town of Clifton. It has a hard sandstone capping of Mesaverde Sandstone and the lower slopes are the easily weathered gumbo clays of the Mancos Shale. Locally the soft hills are called the "adobe hills."

(both in Utah and Colorado) draw on the early name of the river. Even the Grand Canyon of Arizona is part of the story.

Irrigation water from the rivers make it possible to raise world-class fruit orchards in the Grand Valley. Peaches, apples, cherries, pears, apricots, grapes and many other crops come from the area, although cold snaps in the spring make the business rather risky. When the area was first settled, after the Ute Indians were deposed to reservations in Utah in 1880, the towns of Fruita and Appleton became famous for remarkable crops. When farmers irrigated the orchards, salt and other alkali minerals were leached from the good soil and transported downslope. Eventually, Appleton's soil became too salty for good fruit, and most of the valley's orchards gradually shifted to the area around Palisade, where terrace gravel and valley breezes are more favorable for good orchards. Most of Colorado's fruit crop comes from the Delta and Mesa County portions of the Grand Valley.

The "adobe hills" that prevail where there is no irrigation in the Grand Valley, are typical of Mancos exposures in an arid environment. The lower Mancos has calcareous concretions that erode to pimples of one to five meters high. These concretions are common along U.S. Highway 50 between Grand Junction and Delta, and they are identical in origin to the "tepee buttes" of the plains areas near Colorado Springs. The plains buttes are a bit larger, and are in the Pierre Shale, but the Pierre correlates in time and depositional environment with the Mancos.

High quality bituminous coal is abundant in the Bookcliffs. A major uplift in Utah at the end of the Cretaceous provided sediment to fill the Mancos seaway, and the vast

Fig. 208. The Bookcliffs rise from the floor of the Grand Valley to make the north boundary of the valley all the way from Grand Junction to Price, Utah. The basal sands of the Mesaverde form the protective cap that supports the steep, soft Mancos slope. Relief ranges from 400 to 600 meters.

Fig. 209. Lime concretions are abundant near the base and again near the top of the Mancos Shale. This group is along U.S Highway 50 between Grand Junction and Delta.

Fig. 210. A close view of one of the limestone concretions in the lower Mancos. The mound is about one meter across. Shrinkage and swelling of the enclosing clays tends to shatter any fossils or other details in the concretions.

Fig. 211. This coal-fired generator is at Cameo, near the south end of Debeque Canyon. The plant uses Mesaverde coal from a mine just across the Colorado River from Interstate 70. The river is in the foreground here.

Fig. 212. This sliver of the Mesaverde overhangs the power plant at Cameo. Each spring a few more slabs of rock fall onto I-70 in DeBeque Canyon as a result of the weathering in the steep walls of the canyon.

stream system that brought the sand to Colorado also provided a system of swamps and mud to fill the swamps. Lush vegetation buried in the swamps has been converted to low-sulfur coal. By some measurements, only Illinois has more high BTU bituminous coal than Colorado. The coal in the Mesaverde Group (and equivalents) actually extends from Arizona and New Mexico to Montana. Colorado has most of the best coal, but Wyoming and Utah also have valuable reserves. North Dakota, Montana, Wyoming and Colorado also have some younger, Tertiary coal, but the good coking coal is Cretaceous. Sulfur in the Mesaverde coal is about 0.4% compared with the 5% or more that is typical of the Pennsylvanian age coals of the Eastern U.S.

Steep cliffs of Mesaverde sands which overlie the soft Mancos Shale occasionally slide over the Mancos. In 1950 a large chunk of slide material moved near the town of Palisade. About 150 m. of stuff moved, and that would not be too much of a problem except that the piece that moved included a tunnel of the Government Highline Canal, and it was March 8, right at the beginning of the irrigation season for nearly 12,000 hectares (30,000 acres) of orchards and other crops. Repairs were made and a diversion tunnel completed through undisturbed rock in less than two months. The crops were mostly saved! Figure 214 is a photo of another slump that slithered down north of Walker Field Airport at Grand Junction. Other slides and loose boulders that have moved from the Bookcliffs to lie on top of the Mancos Shale provide protection to the soft shale. Rainfall cannot attack the soft shale under protective boulders, and a pedestal rock can develop, such as the "HOODOOS" that are common along the base of the Bookcliffs.

Fig. 213. This chunk of the Mesaverde has slipped down about four meters from its original position. Most slabs tip away from the cliff and disintegrate, somewhat, on the way to the base of the cliff.

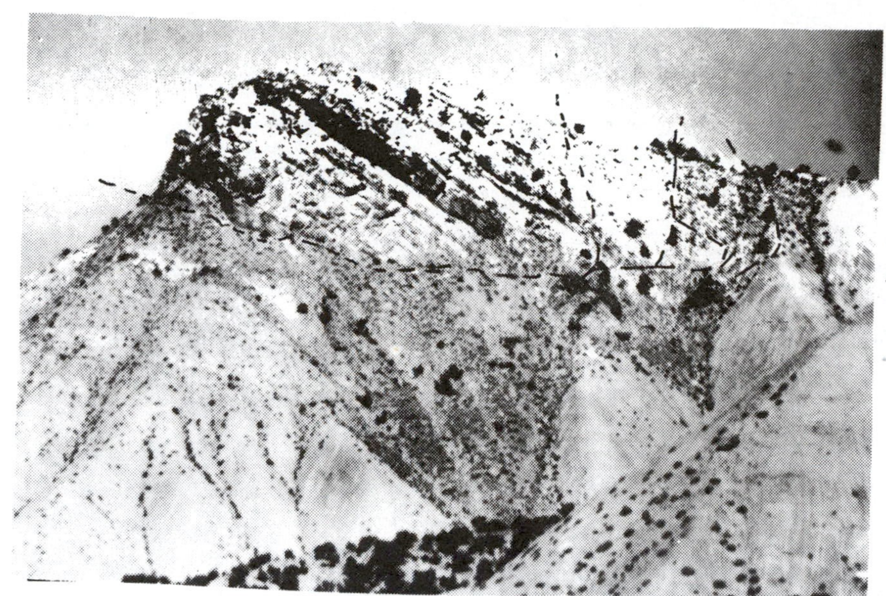

Fig. 214. North of Grand Junction, this portion of the Bookcliffs became detached and glided out over the Mancos. It is displaced several hundred meters.

Fig. 215. When chunks of Mesaverde sands fall and roll or slide out on softer rocks, there is always the opportunity for the loose rocks to become pedestal rocks when the underlying material is protected from the erosion of falling rain. This little family of pedestal rocks (HOODOOS) is at Thompson's Corral and gas station in DeBeque Canyon.

The Uncompahgre Uplift

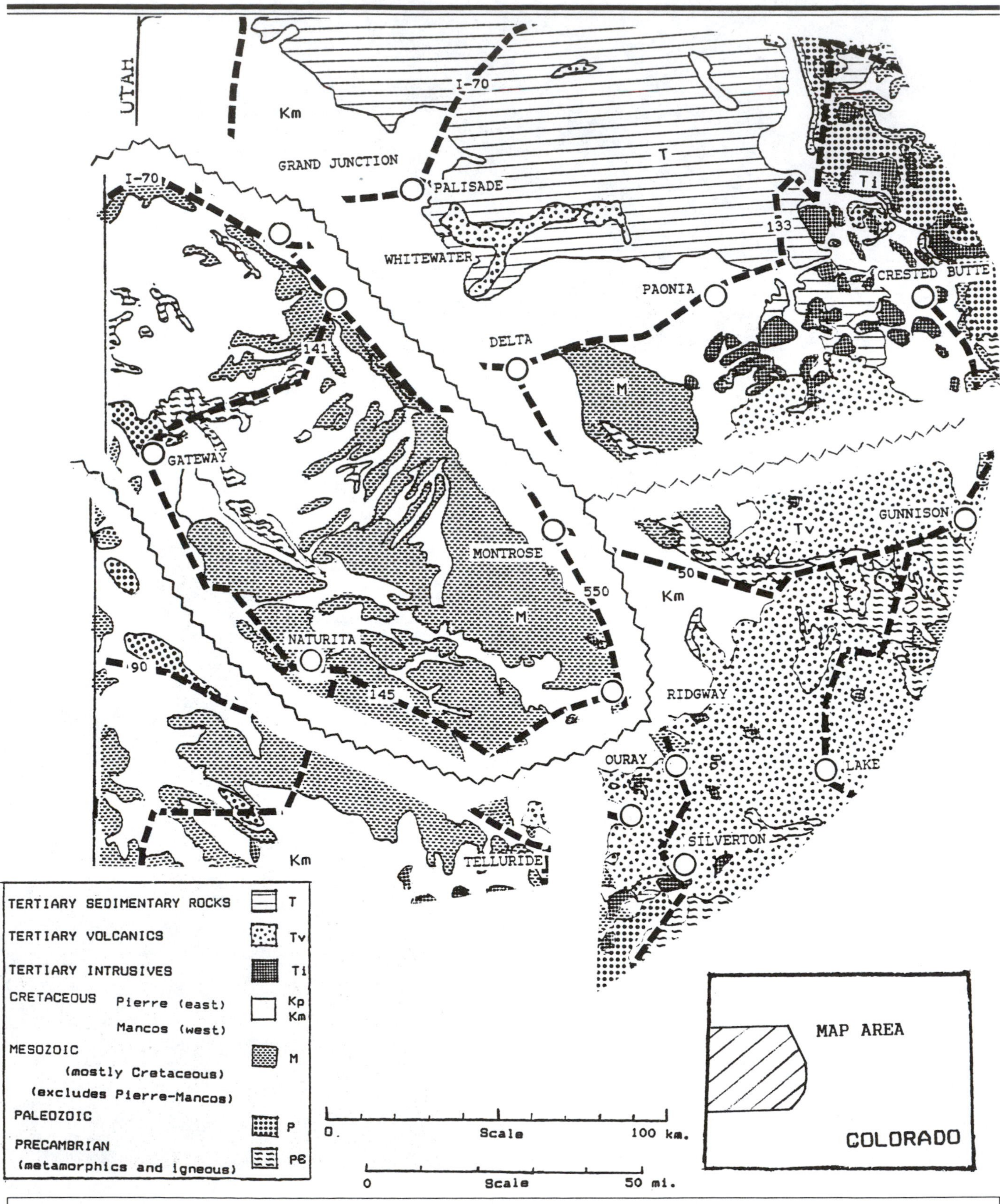

Fig. 216. Location map of the Uncompahgre uplift.

The Uncompahgre uplift, or plateau, is a relatively flat highland approximately 150 km. by 65 km. on the west end of the state. The uplift is bounded on the north and east by the Grand Valley, including the Montrose extension of the valley. On the south the uplift runs against the San Juan Mountains, and the Dolores River and salt valleys of the Paradox basin are on the west side of the uplift. The uplift occupies about 6,800 square kilometers, or about 2.5% of the state. The present geology of the uplift consists of a lump of Precambrian gneisses and schist (Black Canyon Schist), overlain unconformably by Chinle Shale (Triassic). Above the Chinle are the basal units of the stratigraphic section illustrated in Fig. 204. Layers above the Dakota have been stripped off the Uncompahgre by erosion. sediments. Mature sediments are those that are relatively fine grained, well sorted, well rounded and contain only the durable mineral grains that resist the abrasion of long-range transport along streams or coastal beaches. Granite may have 70% feldspar but as the rock is broken and the mineral grains are freed for transport, the feldspars are pounded by pebbles and the harder mineral grains, including quartz. Quartz is harder, has no cleavage and is a little more stable chemically than feldspar. After long transport in an abrasive environment, quartz quickly dominates among the minerals in sandy sediments. Because feldspar is rapidly destroyed, the presence of significant feldspar grains, in sand, tells a geologist that the sediment has only had short transport.

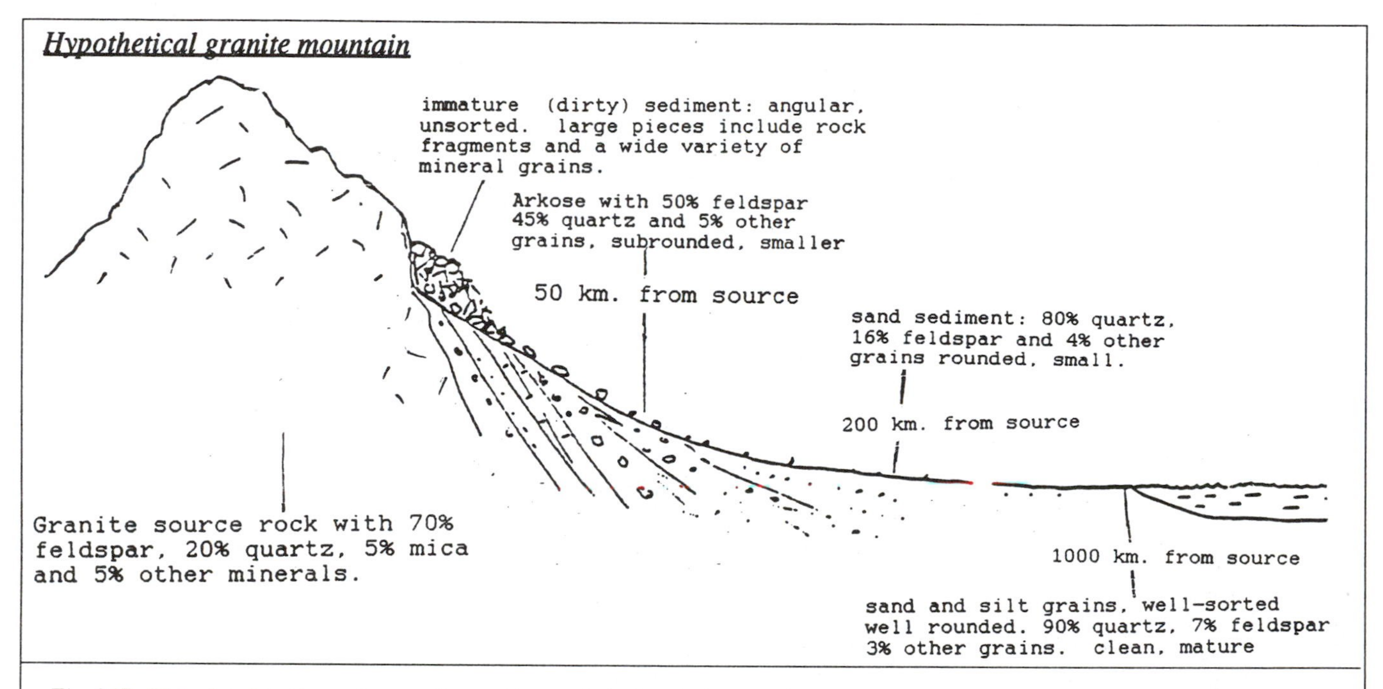

Fig. 217. This sketch indicates how sediments "mature" during the journey from source rocks to depositional basin. Quartz grains resist the wear and tear of transport better than feldspars, mica and most other grains.

The Dakota provides a relatively resistant caprock over much of the uplift.

The present Uncompahgre is a much smaller version of the old Pennsylvanian island of Uncompahgria (Figs. 32 and 33), which was a very important piece of the geologic history of Colorado. Geologists tend to get excited when they find sediments that were stripped from a major uplift of crystalline rock, such as granite or gneisses and schist. The key to an old uplift is the sedimentary rock called "ARKOSE," which is a clastic sedimentary rock with more than 25% feldspar. Most sands and conglomerates have much more quartz and a few other more resistant minerals, instead of the softer feldspars, which are much more abundant in most crystalline source rocks. When the sediments contain a lot of feldspar, geologists know that the material has not been transported very far from the source.

Fig. 217 summarizes the concept of "MATURITY" in

In the basin between Uncompahgria and Frontrangia (Figs. 32 and 33) there is up to 2,000 m. of arkosic fill. The best exposure of this red, continental material is in the 4,200 m. Maroon Peaks, near Aspen, and the name Maroon Formation has been applied to the sequence at that type section. The Maroon Bells are among the highest sedimentary rocks in the state, as most Fourteeners are made of resistant igneous or metamorphic rocks.

The Colorado National Monument is a scenic cliff sequence that forms the north end of the Uncompahgre uplift. Some people familiar with the Canyonlands of Utah and the Grand Canyon of Colorado have criticized the monument as being "just another set of cliffs in the Colorado Plateau region of the American Southwest." But the monument is very accessible and has been properly developed as a national monument. A few kilometers to the west of the current boundary, in the poorly accessible Rattlesnake Canyon, there are 20 or more splendid

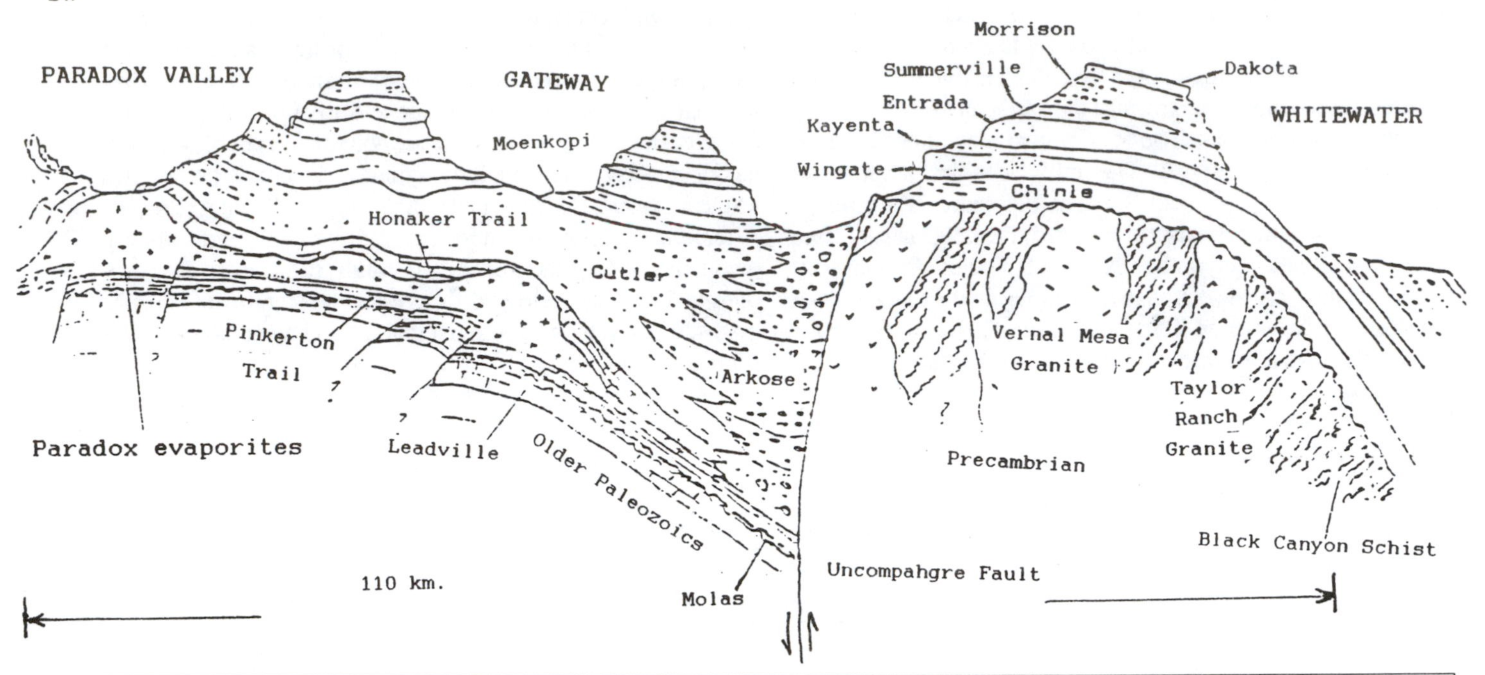

Fig. 218. Simplified crossection of the Uncompahgre Plateau in the vicinity of Gateway and Delta. Vertical scale is greatly exaggerated. The area to the left of the fault is the Paradox Basin.

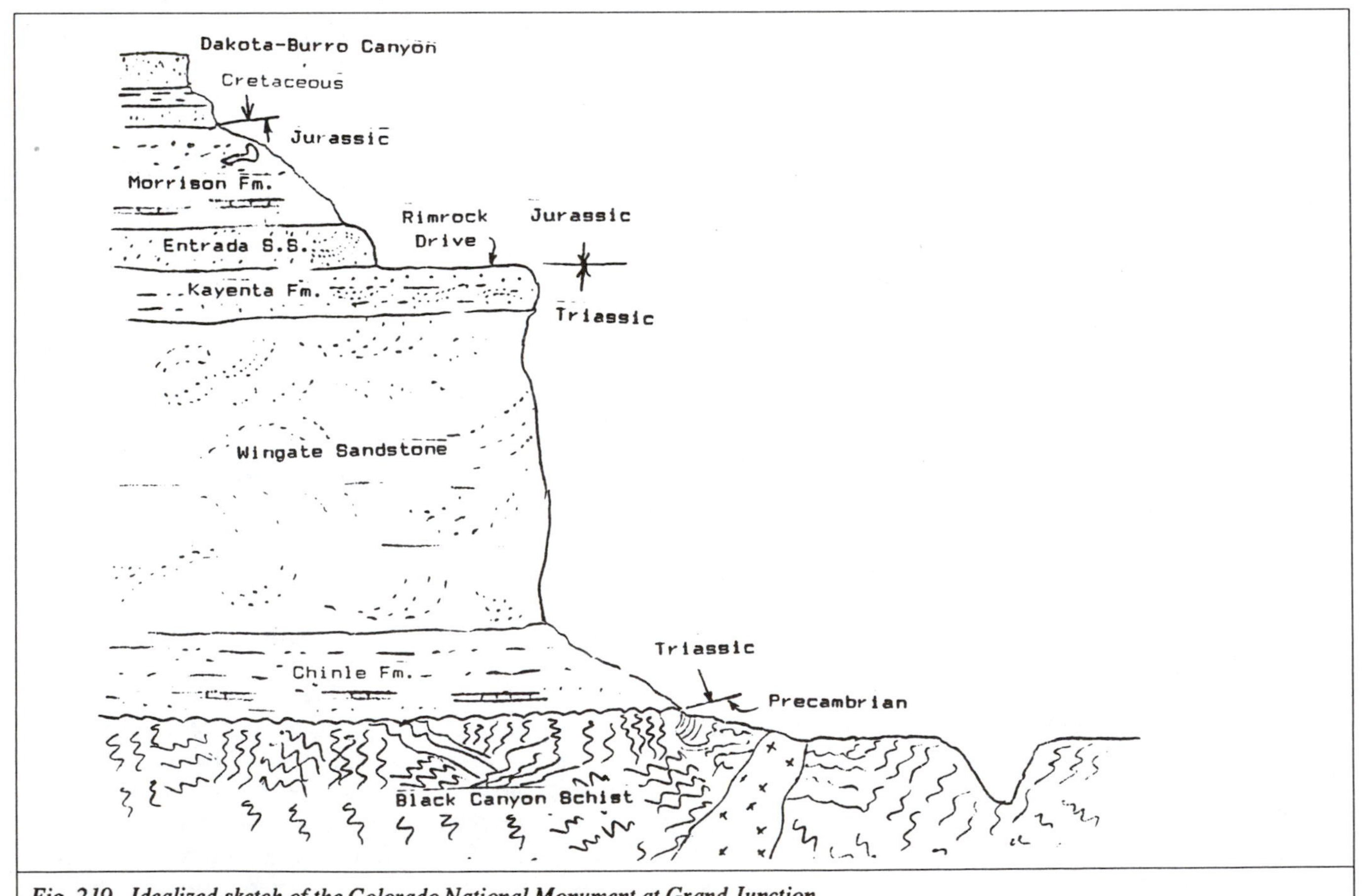

Fig. 219. Idealized sketch of the Colorado National Monument at Grand Junction.

natural bridges and windows. Advocates of the area desire to have Rattlesnake Canyon added to the monument and even elevate it all to national park status, but the bureaucracies involved do not agree so far.

At the monument, the monoclinal fold of the north flank of the Uncompahgre uplift actually separates into a steep fault, pushing the sedimentary sequence on the upthrown block about 200 m. higher than the same units on the downthrown block. The "Redlands" part of Grand Junction gets its name from its view of the red cliffs exposed in the Colorado National Monument.

Independence Rock is an isolated monolith of reddish, Triassic dune sand capped by a protective layer of much harder sandstone with siliceous cement (Figs. 219, 220 and 221). The Wingate Sandstone, with calcareous cement, quickly deteriorates as soon as the protective Kayenta caprock is removed by erosion. The "Coke Ovens" are the remnants of earlier monoliths that have lost their covering, and are rapidly disintegrating as in the simplified sketch.

The colorful Mesozoic sands and shales are in sharp contrast to the dark metamorphic schists and gneisses of the Precambrian Black Canyon Schist that is the underlying "basement rocks" of the Uncompahgre. Deep weathering is typical for most of the top of the Precambrian in the area, suggesting that the ancient, crystalline rocks were exposed at the surface for a long period of time before being covered by red, younger sediments. At many places it is difficult to select the exact spot where weathered metamorphic rock changes to the overlying Chinle Formation. Basal sediments of the Chinle are derived from those very weathered materials, and they are quite similar. Relic dikes criss-cross the weathered Precambrian, sometimes leaving a dim outline with excess quartz and some chalky feldspar crystals to mark an original pegmatite.

Before the National Monument was established, John Otto developed a toll road to the top, and a hiking trail is still maintained along the original 52 switchbacks. The route is called the Serpents Trail.

Just outside the Colorado National Monument is an emerging paleontologic research area. This area has produced a remarkable number of large dinosaur bones, but is rapidly becoming a

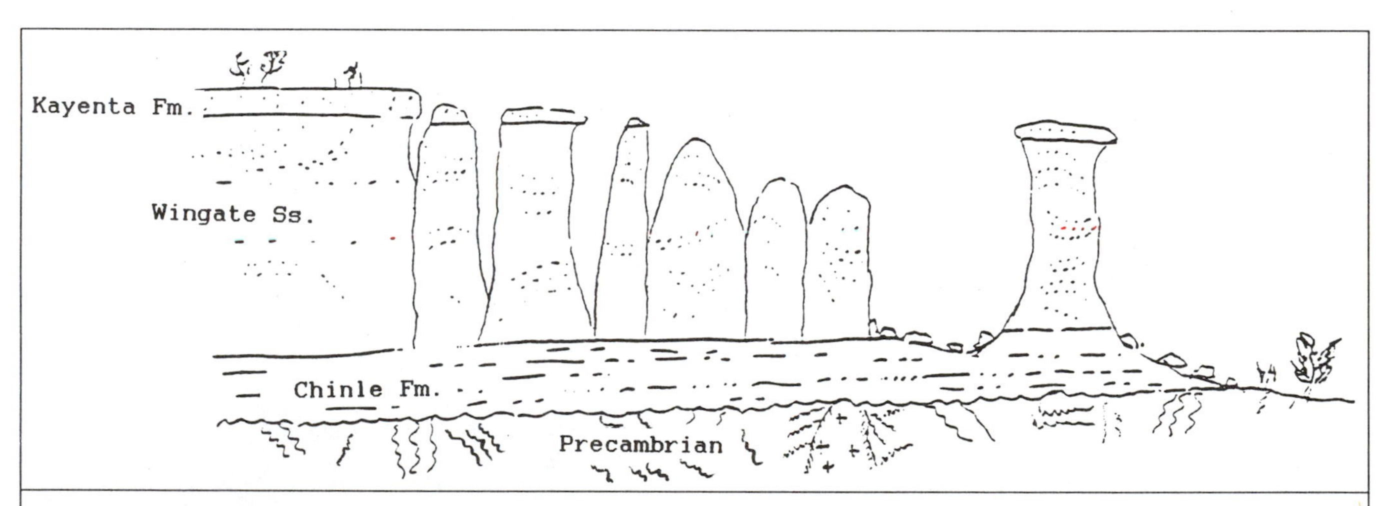

Fig. 220. A sketch to show the erosion process that developed the Coke Ovens and Independence Rock in the Colorado National Monument.

Fig. 221. The Coke Ovens quickly disintegrate when the siliceous Kayenta is gone. The weak, calcareous dune sands of the Wingate form the ovens.

Fig. 222. Fluvial crossbedding is evident in the Kayenta as exposed along Rimrock Drive in the Colorado National Monument.

Fig. 223. Independence Rock is a sliver of Wingate sand with a very small protective cap of Kayenta.

favored site to study early (Jurassic) birds, the earliest snakes, smallest dinosaurs (the size of chickens) and an incredible array of river-dwelling turtles, crocodiles, lizards, plus a variety of plants. Suffocation by ash from far-distant volcanoes may be the key to the demise of so many creatures in one site. This "Fruita Paleontologic Area" is the only parcel of U.S.B.L.M. acreage that is protected for this sort of research (as of 1991). The research area already needs to be expanded to include more, newly-found fossils, and local advocates would like to have these fossil sites added to the Colorado National Monument.

Perhaps the most interesting part of the Uncompahgre, at least to a geologist, is Unaweep Canyon. This deep gash crosses the northern end of the uplift as an enigma. Unaweep is a 1500 m. deep trench in the very resistant Precambrian with no river to do the digging. A small creek drains the east end of the canyon, ending as East Creek. Its counterpart is West Creek, draining the west end. The divide between East Creek and West Creek is 2,100 m. above sea level, but it is in the bottom of a deep canyon. At the divide there is absolutely no stream in the canyon. How did Unaweep Canyon form with no stream to do the scouring?

The two likely origins for Unaweep Canyon are: glaciers, or a river that was diverted elsewhere after carving the canyon. There is plenty of evidence for glaciers in the area, but whether they were of sufficient size and duration to scour such a canyon is doubtful. Glaciers formed on most parts of Colorado more than 2850 m. above sea level with the most extensive glaciers always on the shady north facing slopes of the mountains. At

Fig. 224. Richly colored red cliffs of the Wingate Sandstone tower over Tiara Rado golf course at Grand Junction. This area of the city is call the "Redlands" because of the colorful scenery and the red sands that weather from the Colorado National Monument.

the divide between East Creek and West Creek, the valley floor is about 2100 m. above sea level, and the rims are about 2450 m. above sea level, not high enough for much ice accumulation. Toward the west end of Unaweep Canyon, the rims reach upward to 2,700 m. (9000 ft.), and glaciers did accumulate at the upper end of the side canyons, especially on the shaded south side. Additional ice no doubt spilled in from somewhat higher portions of the Uncompahgre Uplift which extend to the south. Approximately 200 meters of morainal debris is at the west end of Unaweep Canyon, but in this writer's opinion, the ice only modified an already deep and magnificent canyon. The ice was there but did little of the canyon carving because there are only a few patches of possible glacial deposits on the east end of Unaweep. Most of the canyon may have had so much debris clogging the bottom that the ice did not even push out all of the fill, let alone begin much scouring. Nevertheless, the canyon is spectacular, and the cirques and rugged canyons that spill into the lower canyon near the west end, are indeed modified by alpine glaciation.

If glaciers did not carve Unaweep Canyon, what did? It must have been a very large stream. Make it a large river! The Colorado River at Ruby and Westwater Canyons (the north tip of the Uncompahgre) has made a deep gorge, carved partly in the hard Precambrian metamorphics. The Gunnison river has had similar success, even more so, at Black Canyon of the Gunnison. At Black Canyon, however, the Gunnison has had more time to gnaw at the Precambrian. Whatever amount of time that Unaweep Canyon has been abandoned, the Gunnison has been allowed to continue its cutting at Black Canyon. In fact, if the Gunnison was the stream in Unaweep Canyon, it was allowed to accelerate its down-cutting at Black Canyon when it was diverted from the hard channel at Unaweep. The Colorado-Gunnison River (combined) at Grand Junction is only 1,365 m. above sea level, but the divide at Unaweep Canyon is 2,100 m. above sea level. The system is 700 m. lower, in the present channel, than the abandoned divide in Unaweep. That much steepening of the Gunnison's gradient has greatly increased its ability to cut downward in the Black Canyon.

Fig. 225. This view from Rimrock Drive in Colorado National Monument shows the rubble of fallen shafts of Wingate which tumble over the red shales of the soft, red Chinle Shale.

Fig. 226. Between Grand Junction and Fruita, the Redlands Fault has broken the rim of the Uncompaghre uplift, plunging this wedge of Wingate Sandstone about 200 m. into the Grand Valley.

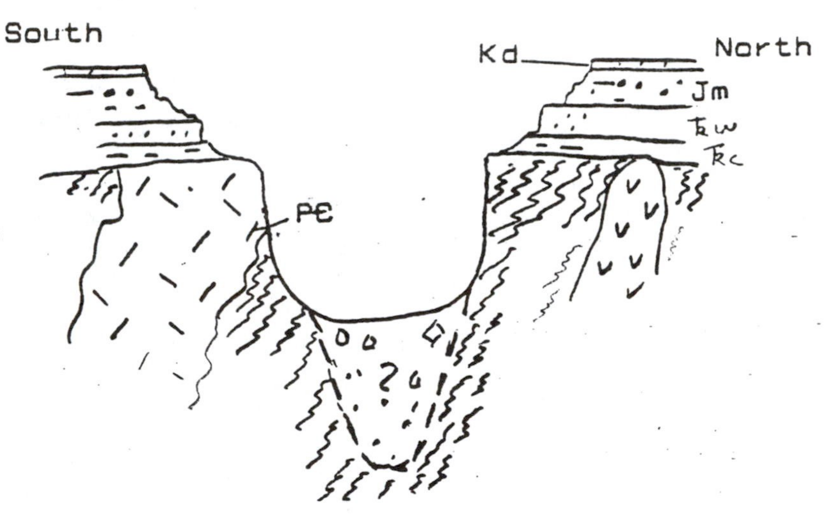

Fig. 227. This simplified crossectional view of Unaweep Canyon shows the "U-shape" that is normally the result of glacial scour on the sides of the valley. In this case, extensive valley fill is probably responsible for most of the shape.

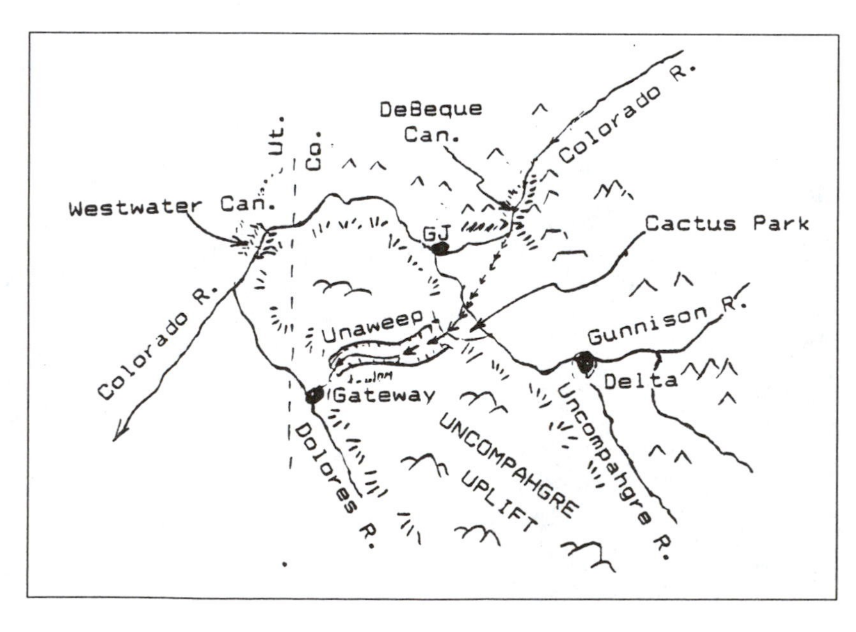

Fig. 228. This map of the streams around Unaweep Canyon shows how the Colorado River was aimed at Unaweep (arrows) before it was diverted across the broad Grand Valley and its underlying Mancos Shale. Gravel unique to the Gunnison River drainage has been found in Unaweep Canyon near Gateway, but distinctive oil shale pieces from the Colorado River have not been identified at Gateway.

Fig. 229. Several working amethyst mines are found in Unaweep Canyon. Shattered rock along minor faults has been mineralized by quartz, flourite, calcite and other minerals. These faceted amethyst crystals are valued at several hundred dollars apiece, especially the two top ones that are the dark "Siberian" shade.

Fig. 228 shows the present pattern of the Gunnison and Colorado Rivers. Note how perfectly the Colorado River is aimed at Unaweep Canyon in its course through DeBeque Canyon. Many geologists believe that the Gunnison River (including Uncompahgre Creek) entered Unaweep Canyon via Cactus Park and joined the Colorado River in Unaweep Canyon, flowing westward to the present Dolores River and on to the present Colorado River valley in Utah.

A few small mineral veins are found in Unaweep Canyon. Some "copper" and "gold" prospects were explored before 1900, but all of these operations were idle by 1912. Only a few carloads of ore were sold. Today there is some activity in a group of amethyst (purple quartz) veins near the east end, and these mines have produced some small, valuable stones classed as "Siberian" amethyst, which is supposed to be prettier purple than just plain purple purple. A granite quarry operated at Taylor Ranch in the 1920's. The site, of course, is the type section of the Taylor Ranch Granite.

South of Unaweep Canyon, and about 40 km. southwest from Delta, Colorado, is the famous Dry Mesa Quarry, where the world's largest carnivorous and herbivorous dinosaurs have been collected. The site was discovered in the 1970's when Eddie and Vivian Jones, of Delta, showed a huge toe bone to Jim Jensen, of the Brigham Young University Earth Science Museum. Jensen's work led to the discovery of the world's largest meat-eating dinosaur, *Torvosaurus*. Additional work at the site found a still larger plant eater that was named *Supersaurus*. Still another plant eater was unearthed that is over 30 m. long and probably weighed as much as a small herd of elephants. The tentative name applied by Jensen is *Ultrasaurus*! These dinosaurs and many others from the Western U.S. are mostly found in the Morrison Formation, of Jurassic age.

Fig. 230. Blocks of Taylor Ranch Granite wait on the rim of Unaweep Canyon. Cut before 1950, the stone was of high quality, but too far from lucrative markets.

Fig. 231. Taylor Ranch Granite is even-grained and of desirable color. The stone was used for several buildings in Denver.

Fig. 232. Dry Mesa quarry on the Uncompaghre uplift can be reached by excellent dirt roads, but local maps should be checked as there are some of the roads in the area that are 4-wheel drive only.

Fig. 233. Dry Mesa quarry is cut from the Morrison, where a variety of species have been recovered, especially some very large dinosaurs. When there is not an active recovery project at the site, the whole quarry is covered to prevent unauthorized digging. Note the lecture platform in the upper right in this 1988 photograph.

Fig. 234. A Brigham Young University graduate student applies a protective plaster cast on a precious bone. Bones deteriorate rapidly when they are exposed at the surface. Bones are shipped, in the casts, and are cut from the cast in the laboratory when assembly and research is done.

The Paradox Basin

In Pennsylvanian time, a large embayment of the sea developed in the Four Corners (the common corner of UT, CO, NM, AZ,) area. Part of the time, the embayment became isolated from the main ocean by some sort of shelf or ridge, allowing the water in the embayment to evaporate. Modern seas have about 3.5% salt, and the Pennsylvanian oceans probably had almost as much. The salt became concentrated as evaporation continued until several hundred feet of salt was deposited in the embayment. This big embayment has been named the Paradox Basin because the salt beds of the basin were first studied at Paradox Valley in Colorado. The salt layers are part of the Paradox Formation, of Pennsylvanian age. Because of the uneven terrain and faults, plus other reasons, there were pockets where the salt was thicker, and places where the overlying material was heavier, causing the salt and accompanying gypsum to squeeze about, much like tooth paste. Several long anticlinal folds developed, and as these uplifts were subjected to greater erosion, the overlying load became even lighter and the doming was exaggerated even more. At Paradox Valley, the salt "dome" exceeds 3000 meters (10,000 ft) in thickness. Salt and gypsum are so soft and soluble that once the top of the dome was breached by erosion, the dome washed away and the structure on the flanks collapsed, leaving a broad valley. Other salt anticlines in the area include Big Gypsum Valley, Sinbad Valley, Professor Valley, Fisher Valley, Salt Anticline (Arches National Monument), Spanish Valley (Moab, UT) and Redmond Valley (Salina, UT). Eagle Valley, east of Glenwood Springs, is cut through an anticline in the gypsiferous Eagle Valley Formation. These beds correlate in time and lithology to the salt layers in the Paradox Basin, and may be another arm of the same seaway.

The name "paradox" comes from the fact that the Dolores River cuts across the Paradox Valley at right angles (Fig. 237) instead of following the trend of the valley — hence a "paradox.". A cluster of Tertiary laccoliths boiled up through the Paradox Basin, disturbing some of the salt domes and forming the La Sal Mountains, which lie mostly inside Utah. Again, salt is the key to the area. The word "LaSal" is Spanish for salt.

The Colorado portion of the Paradox Basin includes an area known loosely as the Uravan mineral belt. During the uranium boom that followed World War II, the Salt Wash member at the base of the Morrison Formation was one of the hot targets for uranium prospecting. Apparently, water seeping through the Morrison leached out the very weak concentrates of uranium from the volcanic ash beds and other sediments of the region. The upper Morrison, the Brushy Basin, is a brightly colored shaly unit with much gray and green volcanic ash mixed with red clays and siltstones. When the uranium-bearing water reached the base of the permeable sands of the Salt Wash Member of the Morrison, the organic matter in the basal sands caused the chemistry of the water to change from an oxidizing to a reducing chemistry and the uranium was deposited in the Salt Wash. Most of the organic material is just flecks of carbon, or small pieces of twigs or bark, but in a few cases entire logs up to 10 meters long were filled with high-grade uranium minerals.

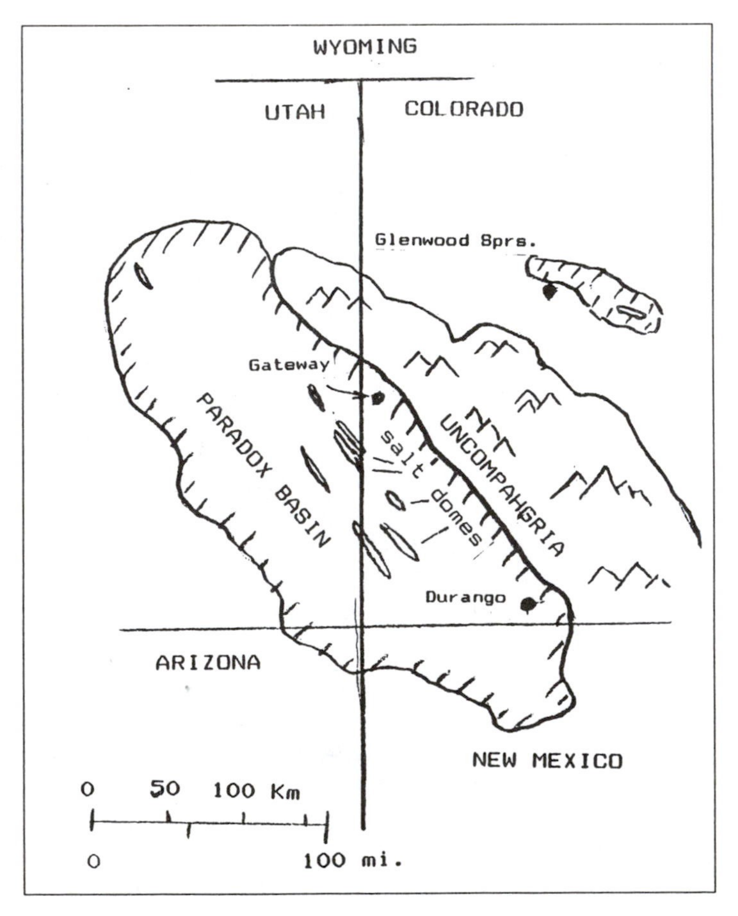

Fig. 235. Map of the Paradox Basin and Uncompahgria, during Pennsylvanian time. Some of the modern salt anticlines are also shown as narrow valleys in the Paradox Basin and also Eagle Valley.

Fig. 236. A photo of a relief map showing Paradox Valley with the Dolores River marked with a heavy line. Direction of stream flow is also indicated. The Uncompahgre uplift is the northwest quarter of the photo.

In the 1950's, Colorado was usually first in vanadium production for the nation, and usually third, after New Mexico and Wyoming, in uranium production. Vanadium is useful in steel making, so the region had a big boom in the wartime 1940's for vanadium mining and concentrating. The uranium boom was mostly after 1945, when the source of power for the atomic bomb was revealed. With cold wars and rapid development of nuclear power plants, western Colorado experienced a wild boom for the uranium industry. The boom ended in about 1980, when the world had too many bombs and the power industry became the target of environmental pressures.

Historically, another boom occurred in the Uravan district. The very name "Ura-Van" comes from uranium and vanadium. The first real boom, however was near 1900 when Marie Curie discovered some medical applications for the very radioactive element radium. Many tons of the high-grade uranium ore can be processed to remove a few grams of radium. At one time, there was a small community in Colorado named Radium, located in the Paradox Valley. The early radium boom ended when someone found richer deposits in Africa, and the Colorado demand stopped abruptly. Today, the town named Radium is located on the Colorado River downstream from Kremmling. The boom for that town lasted only long enough for the buyers to find that the radium "ore" of the area had no radium.

All the booms and busts for radium, vanadium and uranium in the region left an environmental nightmare. Waste products

Fig. 237. Paradox Valley as seen from the southeast end. The low hills in the bottom of the valley include many soft exposures of gypsum. The salt has been dissolved and removed from the surface. The Dolores River is marked with a heavy line and arrow.

Fig. 238. The Gateway vanadium and uranium buying site used during the boom years. This 1988 photo shows how "busted" the boom in the Uravan mineral belt is. Location is on the Dolores River, along Highway 141, about 100 km. southwest of Grand Junction. The houses in the old town of Uravan have also been carted away.

Fig. 239. A Grand Junction morturary faces "minor disruptions" as radioactive mill tailings are removed from a pit about 4 meters deep. Driveways, parking lots, churches, banks, parks, sidewalks, houses and anything else that used "hot" sand in their construction were removed and hauled away in the 1980-93 "remedial action."

Fig. 240. Observers peer down at the Dolores River and the Hanging Flume which clings to the Wingate cliff on the right.

Fig. 241. The remains of the Hanging Flume which once carried water to a gold-bearing gravel terrace high above the Dolores River between Gateway and Uravan. The main cliff is Wingate dune sands and the thinner-bedded sands on top of the cliff are Kayenta.

from all the mining and concentrating left millions of tons of contaminated material scattered over many states. The processed "mill-tailings" are nicely washed and sorted sand, perfect for use in construction and cement-making. A five million cubic meter pile in downtown Grand Junction was a perfect stockpile for contractors for decades. In the early 1990's the pile was moved to a safer location south of town. The job took five years with hundreds of people involved. Several serious injuries and one fatality resulted from the removal projects. The tailings pose a risk to the health of people who live near them. Removing them has been a tragic risk to a few, and horribly expensive. Clean-up of the Uravan concentrators will end up costing nearly half as much as the uranium was worth. The rules will not allow the tailings to be put back in the mines from which they were removed — when they were much more loaded with radioactive elements. Durango, Denver, Rifle and Gunnison also have significant tailings problems.

An unusual coal-fired generator was built at Nucla, and came on stream in 1987. The plant, in western Montrose County, on the west side of the Uncompahgre uplift was modified to be a 110 megawatt "fluidized bed" steam generator. This was the largest unit of its type in the world at startup, and the first in the U.S. of any size. The name "fluidized bed" means that air drafts stir the burning fuel with powdered limestone, keeping the fuel moving as a fluid during combustion. Sulfur in the coal is converted to gypsum as a harmless waste material. The unit uses low grade coal from nearby outcrops of the Dakota Sandstone. Plant capacity is 350 kt. of sub-bituminous coal and 35 kt. of limestone annually.

The Nucla plant is a conversion of an older unit, and the cost was about $880 per kilowatt. A new 200 MW coal-fired plant costs about $1700/KW. The fluidized concept can use very poor quality fuels. Wood chips, combustible municipal waste and other cheap feedstocks have been considered in many areas for this type of plant. Apparently, the Nucla plant could burn an entire month of Montrose County's combustible trash in a very few hours. Mechanical difficulties plagued the operation in the early 1990's.

The Dolores and San Miguel Rivers drain some rich mining country in the San Juan Mountains. Placerville and Telluride are on the San Miguel River, and Rico is on the Dolores drainage. These streams carry placer gold to the Colorado River, and the gravel miners have worked the area vigorously since the late 1800's. As late as 1989, with very low gold prices, placer operations were still sorting the sand, gravel and gold. About 1890, an ambitious aqueduct called the "hanging flume" was built to transport water to a Pleistocene (or Tertiary) gravel terrace high above the Dolores River between Gateway and Uravan. The project was a success, but not enough to produce gold at a profit, so it was abandoned. Pieces of the wooden structure still cling to the Wingate and Kayenta cliffs.

A laminated gypsum deposit occurs at Gateway at the Permian-Triassic boundry. Apparently, a large lake or arm of the sea, became isolated in the area at that critical time in geologic history when most of the fauna and flora worldwide died. The Alabaster Box Mine operated until 1992, cutting 3-inch and 5-inch cores of soft gypsum, which would be turned on lathes to make vases and other exotic sculptures. Old gentleman Chet, operator of the mine and the Alabaster Box Gallery in Grand Junction, died in 1992.

Fig. 242. Placer mines are common the whole length of the Dolores River. The Dolores joins the Colorado River at Dewey Bridge, just across the state line in Utah. At Dewey Bridge there is a large basin before the Colorado River plunges into a swifter stretch of colorful canyons near the La Sal Mountains. The gold that was in the Dolores gravels, as well as the significant gold in the Colorado has accumulated in the big basin at Dewey Bridge. This gravel screen and sluice-box was in operation in 1980, recovering "a few ounces of gold in a few days." Two men could run it. A check with a gold pan recovered 400 gold specks per pan.

Fig. 243. This is the Alabaster Box gypsum mine near Gateway. The laminated gypsum has been cut out with a core drill for sculpturing on a simple lathe.

Fig. 244. A group of "alabaster vases." Alabaster is a fancy name for gypsum, and boosts the price somewhat. The items must be painted with a shiny lacquer, because the gypsum is so soft that it will not take a good polish, and chips very easily.

Fig. 245. Sometines the weathered surface of sandstones becomes pitted. This unusual example is at the confluence of the Dolores and San Miguel Rivers.

Fig. 246. *Two important early cultures took advantage of the geology of the four Corners area. At Mesaverde National Park, top, the "ancient ones" occupied huge caves in the type section of the Mesaverde group of sandstones. At Hovenweep national Monument, below, right, they built their unprotected structures on top of the Dakota Sandstone. The extra hard siliceous cement in the Dakota makes these walls more durable than the softer sands in the Mesaverde buildings.*

Fig. 247. *Another colorful slab of red Wingate Sandstone is about to tumble into the Dolores River Canyon near Uravan. Thin bedded, red, Kayenta sands are on top of the cliff and the tree-covered slope at the base of the cliff is the deep red Chinle Shale. This region of the Colorado Plateau is often referred to as "red rock country."*

The San Juan Mountains

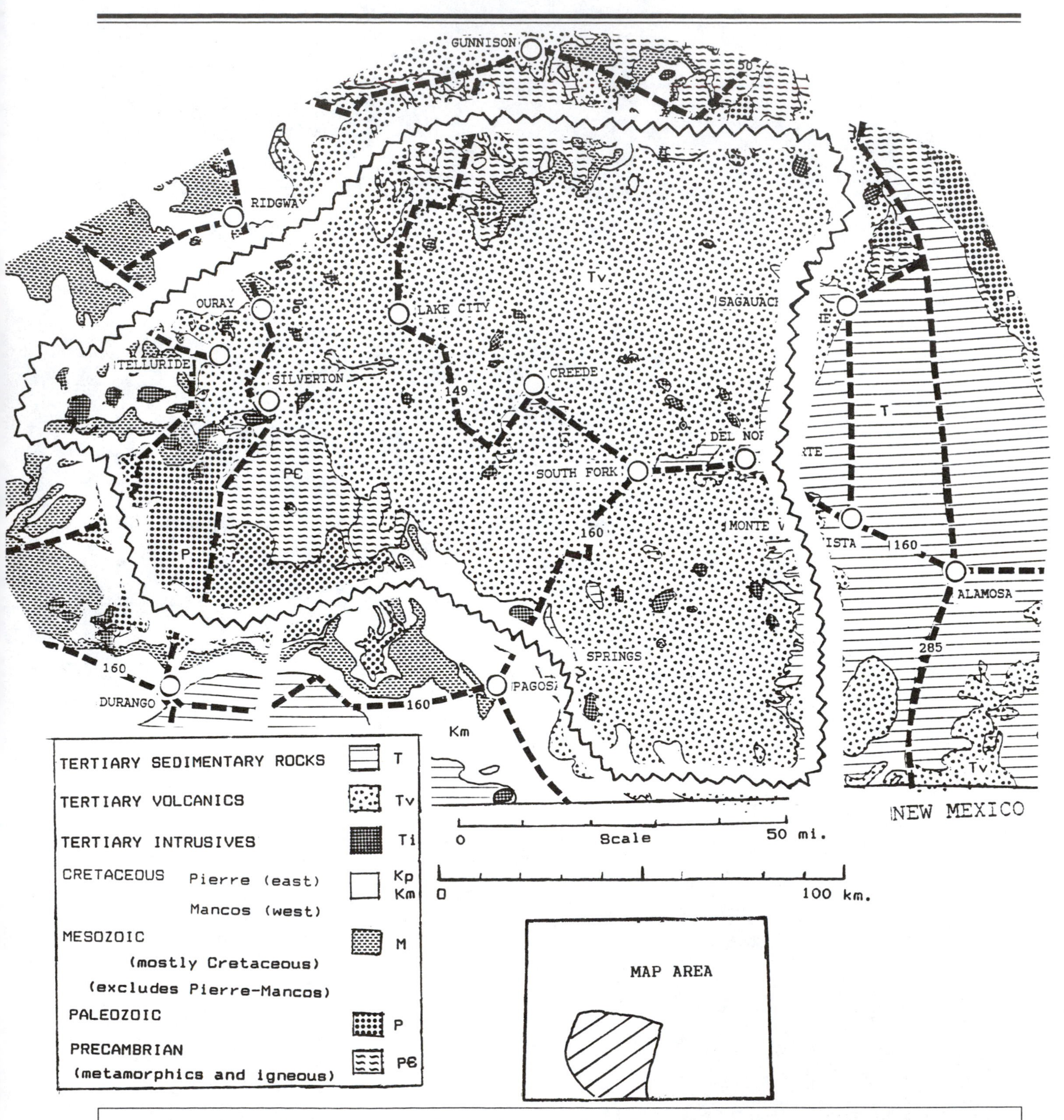

Fig. 248. A geologic map showing the area of the San Juan Mountains. The La Plata Mountains are the southwest end of the San Juans.

The San Juans are the worn-down nubbins of more than a dozen giant volcanoes that blasted thousands of cubic kilometers of volcanic ash, breccia and other rubble over about 15,000 sq. km. (5700 sq. mi.) of the southwestern part of the state during theTertiary Period. Thirteen of those "nubbins" rise above 4200 m. (14,000 ft.). Sticky lava flows were included with the volcanic events, and some portions of the range were injected with fortunes in gold, silver and other minerals. Telluride, Silverton, Creede, Lake City, Ophir, Rico and Powderhorn are a few of the famous gold and silver camps of the San Juan area. Thousands of mines and prospects were dug during the boom years between 1870 and 1920, but today, only a handful of the mines are still operating. Much mineral value remains, but the surface exposures of rich veins are mostly worked out. The next generation of mines in the San Juans will have to find the hidden veins of good ore, or work on a larger scale to take the low-grade material that the technology of 1890 passed over. With more complicated environmental rules, it takes a much better vein to make a commercial mine today than in 1890. If huge pits like the Bingham Mine of Utah could be dug, many areas of the San Juans might have commercial values of gold, silver, zinc, copper and lead. But, such mines are not likely to be permitted in these sensitive times.

Volcanoes come in many shapes and sizes, but there are two end-members that we must understand before we can understand the San Juan Mountains. If the magma is very hot, the erupted lava will be basalt. It will flow freely, stink a great deal, but not explode unless water contacts it. Cooler magmas will produce rhyolite or andesite. So far so good, but the cooler magmas also have an abundance of fluorine, bromine, chlorine, carbon

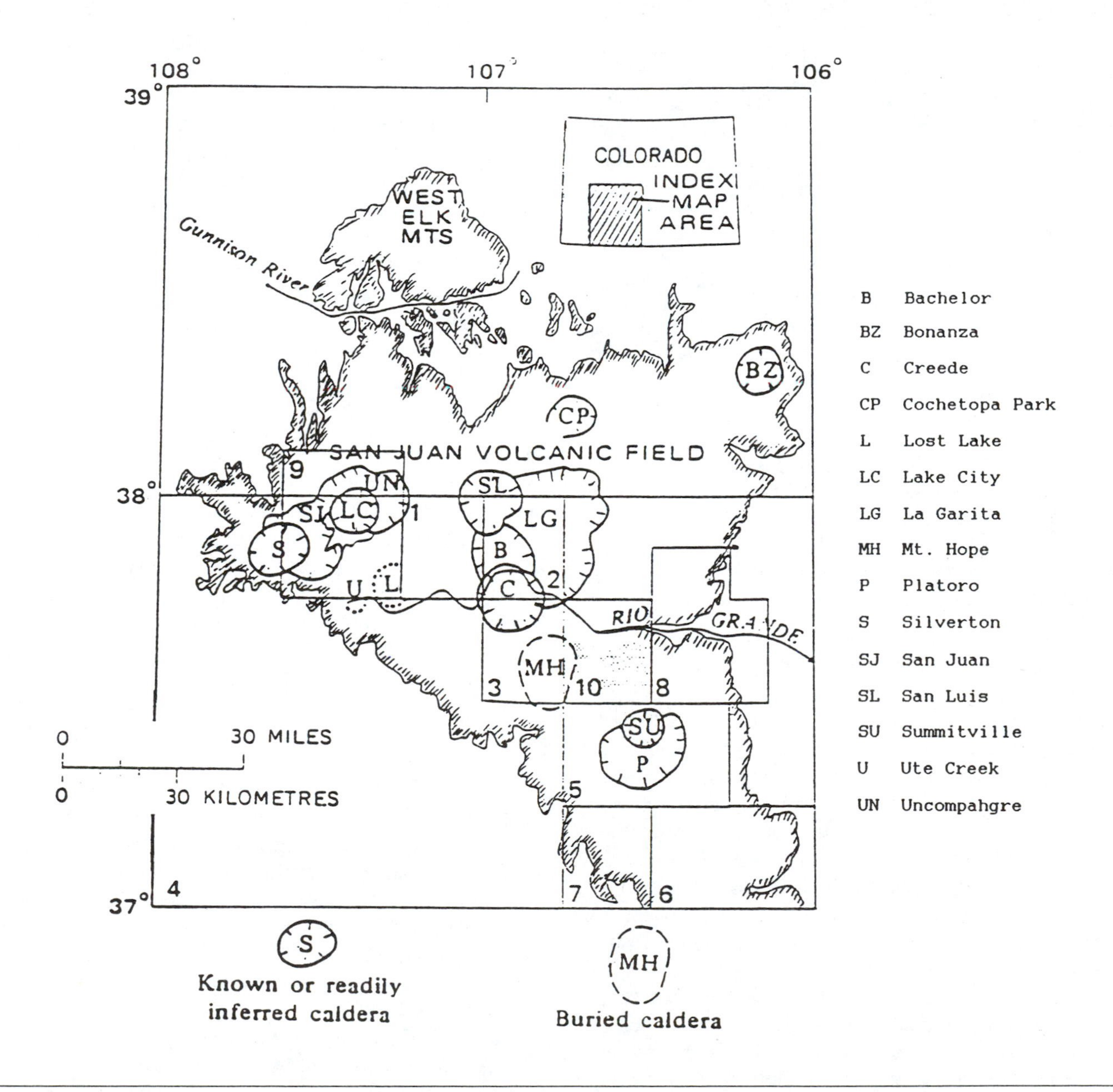

Fig. 249. Map of the San Juan volcanic field, showing 15 recognized calderas. Most of the activity is dated from 19 to 40 million years ago, or middle Tertiary. Basaltic eruptions occur along the eastern margin, adjacent to the San Luis Valley (from Lipman and Steven, 1976, U.S.G.S. Map 1-966).

Fig. 250. Ripple marks seen from the overlook at Bear Creek Falls, on the Million Dollar Highway (U.S. 50) about 5 km. south of Ouray. The ripples were formed in a shallow Precambrian sea. The rock was buried deeply enough to be metamophosed, then a violent, Precambrian uplift pushed the horizontal layers into this vertical position. The whole pile was beveled by erosion before being covered by Paleozoic marine sediments.

dioxide, hydrogen, sulfur oxides, water vapor and, often, many valuable metals such as gold, silver, lead, zinc and copper. The cooler magmas can produce vapors that will form rich veins of the metals named. The cooler magmas are also much more likely to plug up the vent and blast the top off the mountain. Even when the volcano does not blast its top away, the "cool" volcanoes can spout vast quantities of gas, cinders and ash in all directions. The hundreds of tall, beautiful volcanoes around the world are the dangerous ones. Mt. Fuji of Japan, Rainier, Shasta and Hood of the U.S., Mayon of the Philippines and all the other tall, pretty ones were built from countless explosions of ash and clinkers that may blast 10 to 100 km. in any direction. Mt. St.Helens was spectacular, in 1980, and it killed a few people, but the same eruptive sequence is typical of explosive volcanoes. Krakatoa destroyed itself in 1883 and killed 36,000 people in Java. Pelee (north of Venezuela) snuffed out 28,000 with its glowing avalanches in 1902. Tambora's ash darkened the skies of the world for a couple of years when it blew in 1815, and Santorini, in Greece, may have wiped out the entire Minoan civilization 1500 B.C. These are the historic explosive volcanoes that are cousins to the dozens of explosive cones in the San Juans and the West Elk Mountains.

When a volcano blows its top away, or vents away a significant volume from under the summit and the top collapses into the void, a large crater remains. These large craters are called CALDERAS.

Very careful mapping by many specialists has found the remains of at least 15 huge calderas in the San Juans. The volcanoes have been dead for about 20 million years, and tremendous erosion has made the evidence rather subtle, but Figure 249 shows the calderas that are recognized so far. There is a geologic map that includes the Lake City Caldera, and the geologic interpretation of the map describes 55 separate formations that are the deposits left from lava flows, glowing avalanche sequences, tuff-breccias, collapse breccias and other debris from the Lake City source and several of its neighbors.

To many visitors the 5% of the state included in the San Juans is the favorite part of Colorado and, for some, the entire U.S. The town of Ouray (Yoo-RAY) has been dubbed "America's Switzerland" because of the spectacular scenery, reminiscent of the glacially-scoured Alps of Europe. Several geological clues are found in the area of Ouray that help to decipher the history of the San Juans. A dozen other sites might be chosen to unlock the history of the range, but Ouray has great accessibility.

Chapter 1 of the San Juans is found in the Precambrian core that is exposed along the Uncompahgre River and the Animas River drainage, and a few other scattered locations around the range. Limited exposures of Precambrian quartzites, gneiss, and schist intruded by granites and other intrusives tell us that there was a thick accumulation of sediment, probably including marine material, that was buried deeply enough to be metamorphosed and intruded by molten rock. This was all heaved around so that some of the horizontal bedding was twisted into vertical layers which we see today.

Chapter 2 is when all this Precambrian material was worn down and covered by marine limestones, sandstones and other sediments, which include Cambrian, Devonian and Mississippian rocks. This chapter has some holes in it because there are no "pages" of Silurian or Ordovician history in the book for the San Juans.

At Molas Pass, between Ouray and Durango, the Mississippian Leadville Limestone was uplifted gently and became a KARST terrain. The word "karst" means a limestone terrain with sinkholes, caverns and other features that form when limestone is dissolved by surface weathering. The soil that accumulated on the surface and in the sinkholes appears today as a bright red, hematite-rich formation called the Molas Formation. Many other parts of western Colorado have the red Molas atop the Leadville. It is usually less than 15 m. thick.

The uplift of the Uncompahgre during Pennsylvanian time affected the western end of the San Juan area. Red sediments (including some arkose) mixed with marine limestones of the Hermosa Group between Ouray and Durango. The more serious Pennsylvanian uplifts removed the Paleozoic sediments from the core of the San Juans. By late Cretaceous time the area

Fig. 251. View to the north over the town of Ouray. The glacially-scoured valley has red Paleozoic sediments surrounding the town, and these are capped by nearly a thousand meters of volcanic breccia. A yellow-stained stock seen on the east wall of the canyon is called the "blowout" and is pock-marked with old gold mines and prospect holes. The scene is from the upthrown block of a major fault that dropped Pennsylvanian rocks against vertical Precambrian quartzite layers.

finally subsided again to receive the Mancos-Pierre seaway.

The first hints of the Laramide Orogeny were some igneous snorts, including a small stock called the "blowout" at Ouray. This gray, 70 million year old feature roasted the surrounding rocks and apparently loaded the surroundings with pyrite. When the pyrite weathered, it changed to yellow and brown limonite. Gold, silver, lead, zinc and copper are also associated with this stock at Ouray, and they are the beginnings of exciting mineralization in the area which would continue for the next 50 million years.

Probably before Oligocene time the west end of the area bobbed upward and shed a thick apron of gravel that became the Telluride Conglomerate. Because this jumble of stones, which can be 350 m. thick, is made of a variety of rock types, and has a loose matrix, it becomes a target for mineralization. When hot, metal-bearing solutions entered the area in association with Tertiary stocks and volcanics, the variety of stuff in the Telluride Conglomerate neutralized the typically acid vapors of the intrusions, and provided an ideal site for rich veins of metals. Good veins have formed in all kinds of rocks in the San Juans, but limestones are preferred for the necessary chemistry. When limestones are not available, a conglomerate with limestone fragments is the next best choice. In the San Juans there are thousands of meters of volcanic debris which are highly mineralized in some places. There was so much rich, mineralizing material coming up in the area, it had to go someplace, so even the volcanics can become high-grade ore. Recent studies in the San Juans indicate that some of the richest deposits, in the Creede and Lake City areas for example, were formed in part by surface waters reworking the low-grade metal deposits and enriching them by recycling. Even gold, which is almost impossible to dissolve in ordinary acids, can be mobilized if the particles are extremely fine and there is plenty of time for the weak solutions to work on the gold.

Fig. 252. At Box Canyon, in south Ouray, Devonian marine sediments lie unconformably on top of the vertical Precambrian quartzites.

The end of the San Juan book includes some additional late Tertiary and Quaternary uplift that forced the intense erosion that has sculptured the present range. Ice-age scouring was the finishing touch for a magnificent set of jagged peaks, including

Fig. 253. A fence guards the trail into Box Canyon at Ouray. The sheer wall on the left contains scrubbed rock, breccia and slickensides, and is where the Cambrian through lower Pennsylvanian strata slid downward under Ouray. Here horizontal, red Pennsylvanian beds (Cutler Group) on the right, are in contact along the fault surface with vertical, Precambrian quartzites on the left.

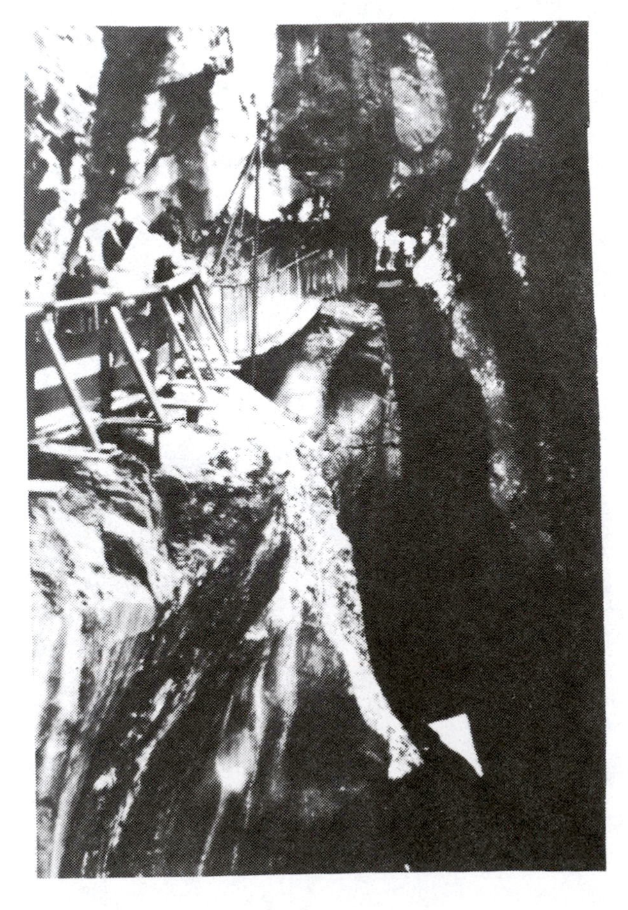

Fig. 254. Visitors peer into Box Canyon from a catwalk built into the gorge. At the end of the trail, the creek thunders past the observers, spraying them, before splashing into the dark chasm below.

Fig. 255. The Million Dollar highway is U.S. 550, heading south out of Ouray. Waste gold ore was used for gravel in the original roadbed, but now snow removal, normal maintenance and lawsuits over accidents are all in the million dollar range. The scenery is also priceless. The view here is of steeply -ilted slate, and quartzites. High above these rocks is a sequence of as much as a thousand meters of volcanic sediments. Glaciers scoured everything below the volcanics.

This is a summary of the history of the San Juans:

- a. *thick accumulation of Precambrian sediments, enough for metamorphism.*
- b. *intrusions, folding, and much uplift (Precambrian).*
- c. *erosion to remove miles of overlying rock (Precambrian).*
- d. *Cambrian seaway advances, leaving sands of Ignacio Fm.*
- e. *Pre-Devonian erosion.*
- f. *Devonian (Ouray) and Mississippian (Leadville) deposits.*
- g. *minor uplift allows karst to develop on top of Leadville.*
- h. *Uncompahgria surges up, shedding red sediments on flanks.*
- i. *Mancos sea covers area, with Dakota sand at base.*
- j. *scattered intrusions at the close of the Cretaceous.*
- k. *abrupt uplift of west end sheds Telluride Congl. (Tertiary).*
- l. *world-class volcanoes blast area until 19 mil. years ago.*
- m. *basaltic lavas on east end, ending about 4 mil. years ago.*
- n. *erosion removes 2-5 km. of material from most of the range.*

---

a dozen of the 4,200 m. variety (fourteeners).

If we return to Ouray, there are some special things to see there. About 11 km. north of Ouray is the town of Ridgway, and U.S. Highway 550 cuts the edge of a terminal moraine there. A massive glacier scoured the upper Uncompahgre River drainage and dumped a pile of unsorted, unrounded, unstratified material there that effectively dammed the valley, and made a large lake. The moraine dam is actually two different moraines, with an older, and higher one, appearing a few kilometers farther downstream (north). The lower, and more southerly moraine rises about 100 m. above the level of the lake basin. It is being exploited for building lots and homes. The older and higher moraine rises more than 500 m. and has evergreens on it to mark its much greater altitude than the lesser moraine.

Ouray has a rash of hot springs, ranging in size from the big one at the municipal pool north of town to smaller ones in the motels, such as the grotto-like sauna in the sub-basement of the Weisbaden. Skinny dipping pools are used along the creeks, even in winter. These hot springs tell the geologist that there is open access to deep, geothermal heat along faults. On the south end of town the road zig zags its way over an abrupt escarpment of hard, vertically-bedded, Precambrian quartzite. The town is in alluvial stuff on Paleozoic redbeds of the Pennsylvanian. The escarpment has a sheer wall of Precambrian in fault-contact with the Pennsylvanian. At the escarpment there is a deep gorge with a waterfall hidden at the end in a place exploited by local service groups as "Box Canyon Falls." With hand rails and well-built trails, visitors can go to the head of the falls and feel the gorge vibrate with a major tributary of the Uncompahgre River roaring past one's face and plunging into a dimly lit pool below. Another trail includes a 100-step staircase and a view of a dramatic angular unconformity, where vertically-wrenched Precambrian marine quartzites are capped by Devonian marine limestones and silts. From the gazebo along the upper trail, one can see "the Blowout," which is a 70-million year old stock that roasted its way into the area and injected minerals that have now rusted to form colorful yellow stains to mark the veins. These yellow, rotted veins are called gossans, and are described in detail on pages 72 and 73. Above the brightly stained Paleozoic sedimentary rocks (Cutler Group), the unstained San Juan volcanics lie as somber gray cliffs on top of the section in most parts of the San Juans. At Ouray the San Juan volcanics (tuffs and breccia) are unstained because they were deposited after the mineralization of "the Blowout," which was the very end of the Cretaceous. Most of the mineralization of the San Juans however, is middle Tertiary.

Upstream from Ouray, U.S. 550 becomes the "Million Dollar Highway." The roadway has been chiseled from the

Fig. 256. A small, gabbro dike on the east side of the Uncompahgre Valley between Ridgway and Ouray. The dike can be seen from U.S. 550 which is shown on the left side of this photo. The open valley here was full of water when the Ridgway moraine dammed the river.

Fig. 257. Bear Creek Falls plunge from a small hanging valley, about 5 km. south of Ouray. The rock alongside the falls is Precambrian slate.

Fig. 258. A beveled exposure of hard slate can be seen at the Bear Creek Falls roadside stop. Glacier striations (scratches) and a little polish are obvious. The unsorted, unstratified, mostly unrounded material that covers the exposure are part of the lateral moraine of the glacier from the main valley.

Fig. 259. This view is looking south from the terminal moraine of the glacier that scoured the Uncompahgre River valley. The till is late Wisconsin, and is covered with sagebrush and scattered junipers. Earlier moraines are much higher, pushed a bit farther downstream, and are covered mostly with junipers and pinyon trees. The flat floor of the valley in this view was flooded by a moraine-dammed lake until the stream was able to breach the dam and drain the lake. The town of Ridgway is just out of view to the right.

Fig. 260. This road cut on U.S. 550, just north of Ridgway, shows the unsorted till of the terminal moraine. Rocks in the till include red sandstones, slate, quartzite, volcanics, intrusives and virtually everything that is exposed in the whole river valley.

Fig. 261. About 18 km. north of Ridgway, and 1 km. north of the Ridgway Dam is this fine-grained diorite sill, exposed on the west side of U.S. 550. Dakota sands cap the hill, so the sill squeezes into the Morrison, roasting both the top and basal contacts. The base of this sill has some very unusual stripes that are presumably caused by repeated surges of new material into the sill. Each surge allowed a layer of heavy minerals to settle to the bottom before the next surge brought in a new layer of undifferentiated magma.

Precambrian quartzites and slates, and also the overlying volcanics, by sheer stubbornness. Wintertime avalanches in the area often close this route, and a snowshed has been built to protect travelers (and snowplow drivers) from the hazard. Spring thaws and summer thundershowers pepper the road with rocks that are constantly loosened along the route. The million dollar idea comes from the fact that gold mines supplied some of the gravels used in the original road. It also cost more than a million of those dollars of the 1880's to make the road. Nowadays, cleanup and snow removal probably cost a million dollars most years.

At Bear Creek Falls, a scant 5 km. upstream from the Box Canyon exit, is an overlook where ripple marks can be seen in the vertical Precambrian quartzites, the water falls about a hundred meters over a slate cliff, and glacially-scoured slates underlie a patch of unstratified, unsorted moraine. Cliffs of crudely stratified San Juan tuffs and other breccias tower above to about 3,600 m. above sea level.

Fig. 262. A car ducks into the short avalanche tunnel on the Million Dollar Highway. The tunnel was built in the 1980's but an avalanche in the winter of 1991-92 was wider than the tunnel and caught a snowplow outside. One of the crew was lost, one was rescued. Plans are to lengthen the tunnel. Precambrian metamorphics are exposed here. The San Juan tuffs are much higher.

Fig. 263. Near the tunnel shown in figure 262, is this monument to snowplow drivers lost in separate years in the 1970's. Another monument is to a family, all lost in the same avalanche chute where the short tunnel is now.

The Lake Fork of the Gunnison River is the focus of a lot of geology. There is a world class gold camp in an obscure Tertiary caldera, and the namesake lake, Lake San Cristobal was dammed by the famous Slumgullion landslide. From the slide can be seen several Fourteeners, one of them a classic horn named Wetterhorn Peak. A short drive over Slumgullion Pass takes one to the Continental Divide, and glacial features abound in the area. Hunting and fishing are among the best in the nation, but one old codger defied a long, cold winter by eating all his companions!

Fig. 264. A memorial to avalanche victims in the Telluride cemetery. Avalanches take many lives each year in Colorado.

Fig. 265. Travelers in the early summer examine debris left from a powerful avalanche that ran during the winter near Ophir. The pile of small trees indicates that the avalanche clipped trees in an area that had been spared for about 20 years.

Fig. 266. Lake San Cristobal is in the upper section of Lake Fork of the Gunnison River. The area here is intensely glaciated, making this a very scenic valley. The lake is dammed partly by the Slumgullion landslide. View is from the Slumgullion pass road.

Fig. 267. Looking downstream on Lake San Cristobal. Yellow muds of the Slumgullion slide enter the lake from the right, and waves slosh the material loose and spread it around the lake during windstorms. Weathered gold veins in the area resemble this limonite-colored muck, and the slumgullion slide probably contains some very fine gold.

Fig. 268. The San Cristobal Whale! A roche moutonne that is surrounded by water is called a whaleback. This one was scoured by the glaciers moving from right to left, plucking away at the steeper, downstream side. And, Alfred Packer's party starved at the sight of this big "fish."

Fig. 269. Slumgullion slide continues to creep toward Lake San Cristobal from near the top of Cannibal Ridge.

Fig. 270. A closer view of the Slumgullion slide shows the "drunken forest" typical of unstable slopes that are still moving.

Fig. 271. View is across Lake San Cristobal from the Slumgullion slide. The light-colored piles near the top of the photo are mine waste piles and the mines were active in the 1980's. The gully that contains these gold mines has supplied sediment that spreads out into Lake San Cristobal in the bottom of the photo. The lower right corner of the photo shows Slumgullion slide material entering the lake from the near side.

Fig. 272. *This vein of limonite is exposed in a roadcut on the Slumgullion Pass road. Although the vein is not available for mining, it probably contains a little gold.*

Fig. 273. Monument to Alferd Packer's five-course meal. It is alongside the Slumgullion Pass road, just after the road crosses Lake Fork at Lake San Cristobal.

Fig. 274. Uncompahgre Peak is the fourteener on the right. The three peaks on the left are Broken Hill, Wetterhorn and Matterhorn. All are over 3960 m. (13,200 ft.). Of course, all are true glacial horns. The view is from the Slumgullion Pass road.

Fig. 275. Near the west end of the San Juans is Mt. Sneffels, a fourteener made of intrusive igneous rock--perhaps the neck of a much higher volcano. High peaks on each side of Mt. Sneffels are made of horizontally-layered breccia and volcanic tuff. At least one kilometer of material has been removed from all the San Juan peaks.

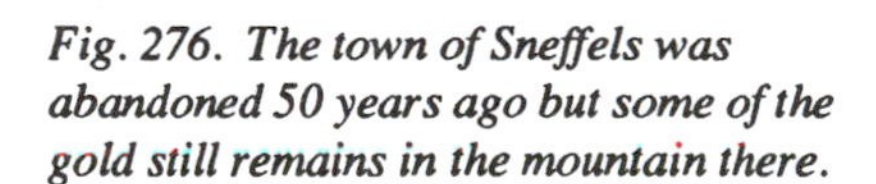

Fig. 276. The town of Sneffels was abandoned 50 years ago but some of the gold still remains in the mountain there.

Fig. 277. Much of the high San Juan mountains is accessible only by 4-wheel drive vehicles. There is no need to go as fast as this one, which has a front wheel airborne, on Imogene Pass, which connects Ouray with Telluride.

Fig. 278. View to the north, down Canyon Creek, which drops over Box Canyon Falls. The foreground has piles of mine tailings from the Camp Bird mine. Most of the cliff on the left is the Telluride Conglomerate. The top half of the mountains in the background is the San Juan Fm. (tuffs and breccia). The Camp Bird produces from the Argentine vein, which is a throughgoing shatter zone that is intermittently traced from Silverton to Sneffels.

Fig. 279. A rock glacier in Imogene basin. This huge stream of loose rock (talus) creeps out of the cirque at the head of Imogene Creek. Rock glaciers include much ice a few meters below the surface which is really permafrost. The ice is insulated from the heat of summer by the overlying boulders. A rock glacier therefore acts a lot like a glacier that is saturated with boulders.

Fig. 280. The toe of the Imogene rock glacier advances toward the mine dumps of the upper Camp Bird Mine.

Fig. 281. Telluride as viewed from the top of Bridal Veil Falls. The town sits in a colorful U-shaped valley with red Cutler beds overlain with the Telluride Conglomerate. About timberline the tuffs and breccias of the San Juan volcanics appear. Old mine tailings ponds are perched upstream from the modern town of Telluride. Skiers and a tourist economy have supplanted the old mining economy of the region.

Fig. 282. A boulder of Telluride Conglomerate. The matrix and some of the clasts are replaced with mineralization. A rind of replacement shows on some of the larger cobbles. This rock type is especially suited for replacemnt by rich, metal-bearing solutions.

Fig. 283. Bridal Veil Falls spill over the type section of the Telluride Conglomerate at Telluride. The unit here is about 155 m. thick. An old power plant clings to the top of the cliff in the upper right corner of this photo. The plant is being preserved for special historic value.

Fig. 284. Molas Lake is the type section of the Molas Formation, which is mostly soils and red, oxidized material left on the top of the Mississippian Leadville Limestone. Euolus, Windom and Sunlight peaks are the fourteeners here as seen from U.S. 550.

Fig. 285. Colorful red arkoses and fossiliferous marine carbonates of the Hermosa Group (Cutler) lie at the top of Molas Pass at 3273 meters above sea level (10,910 ft.). Volcanics cap the section, mostly above timberline here.

Fig. 286. On Cimarron Creek these pinnacles of San Juan volcanics show the typical weathering habit of these layers of tuff, ash and breccia that were blown from a number of volcanoes in the San Juans.

Fig. 287. Chimney Rock is a remnant of the thousands of meters of ash and breccia deposited on top of the San Juan volcanic pile. The layers of individual blasts are clearly seen in this unusual column exposed at Owl Pass, on a dirt road leading from Ridgway to the Cimarron basin.

Fig. 288. An abandoned ore concentrator at Silver Lake lies above 3600 m. in the cirque at the head of Arrastra Gulch. Miners hoped for a lot of snow when they worked these high deposits in the winter because they could move from work to dinner and visit the camp followers all through a network of snow tunnels. With less snow, all movement was done in full exposure to the biting winds.

Fig. 289. Arrastra Gulch leads from Silver Lake down to the Animas River, about 5 km. upstream from Silverton. The area is near the center of the Silverton district which was a major focus of the gold, silver, lead and zinc mining from 1870 to the early 1900's. Mine tailings are seen in the bottom of the Animas Valley shown here. Most of the rocks are volcanics in the Silverton caldera.

Fig. 290. This sharply-tilted sandy member in the Mancos Shale gives an approximate earliest possible date for the uplift of the San Juan platform before the volcanoes spouted off. The location is along U.S. 160 between Durango and Pagosa Springs. The Mesaverde is above this unit and it is also forced up like this.

The east end of the San Juans is a transition area. The wild explosive vents and calderas of the west end must yield to the late Tertiary basaltic flows on the east end that border San Luis Valley. Igneous rocks of the San Luis rift valley boil up from the deep mantle, and are ultramafics. San Juan volcanics are intermediate to felsic, and were supposed to erupt violently. They were like Mt. St. Helens, Krakatoa, Mayon and Pinatubo of the Philippines, Unsen and Fuji of Japan—in fact, all of them all popping off together for 20 million years from about 20 to 40 million years ago. The San Luis rift zone gets its own chapter, but the eruptions were more like the quiet Hawaiian volcanoes, where tourists rush to the rim of the crater to watch the fountains of basaltic magma. Iceland, the Snake River Plains, of Idaho, Washington's Columbia Basin and even the Grand Mesa are the kinds of features that develop from rift eruptives, not strato-volcanoes and peaks that disintegrate to become 20-km wide calderas. Tourists must flee from these!

How could a beginner in geology figure out when the mountain-building of the San Juans started? Some clues are available along Route 160, between Durango and Pagosa Springs. The San Juan Basin, south of the San Juans, is an oil-rich region of Colorado and New Mexico, occupying thousands of square kilometers. The Mesaverde Group is late Cretaceous age (trust the experts) and it forms part of the rim of that basin. It was folded, as part of the San Juan uplift, after it was deposited. Therefore, the San Juans started up after all the Cretaceous sediments were in place. The uplift was extensive, as indicated by Fig. 290 which shows a sand unit in the Mancos that was wrenched up at an angle of 50 degrees. The Mancos seaway covered all of the San Juan area, and all those soft sediments, plus the Mesaverde were eroded off the top before the volcanics came blasting out.

Fig. 291. Pagosa Springs is a group of hot water vents that gush up from carbonate-rich sediments on the south flank of the San Juans. The hot water has been piped into some public buildings for heating. The water is rich in dissolved minerals, giving a rather strong sulfurous odor to the springs. At the surface, some of the water evaporates and it must deposit some of the dissolved carbonate. The result is travertine cones and terraces.

Fig. 292. At North Pass, between Saguache and Gunnison along Route 114, this cluster of columnar joints is exposed in the road-cut. Note that the columns are not a precise crystal shape with consistent angles. A close look, below, shows that the columns vary from 3-sided to seven sides. The rock is of intermediate composition.

Fig. 293. Layers of volcanic breccia and white tuffs exposed in the eastern San Juans, and viewed from the road toward Wolf Creek Pass from Pagosa Springs (U. S. 160). Glaciers scoured the valley to make it U-shaped, with mounds of morainal till in the bottom.

Fig. 294. Coarse, angular breccia exposed at a turnout on U. S. 160, south of Wolf Creek Pass.

Fig. 295. Columnar joints that lie horizontal reveal a feeder dike along the Wolf Creek Pass ascent on the south side.

Fig. 296. Four separate lava flows can be picked out of the road cut south of Wolf Creek Pass.

Fig. 297. Normal weathering and "mass wasting" processes begin to cover an abandoned railroad near South Fork, in the eastern San Juans.

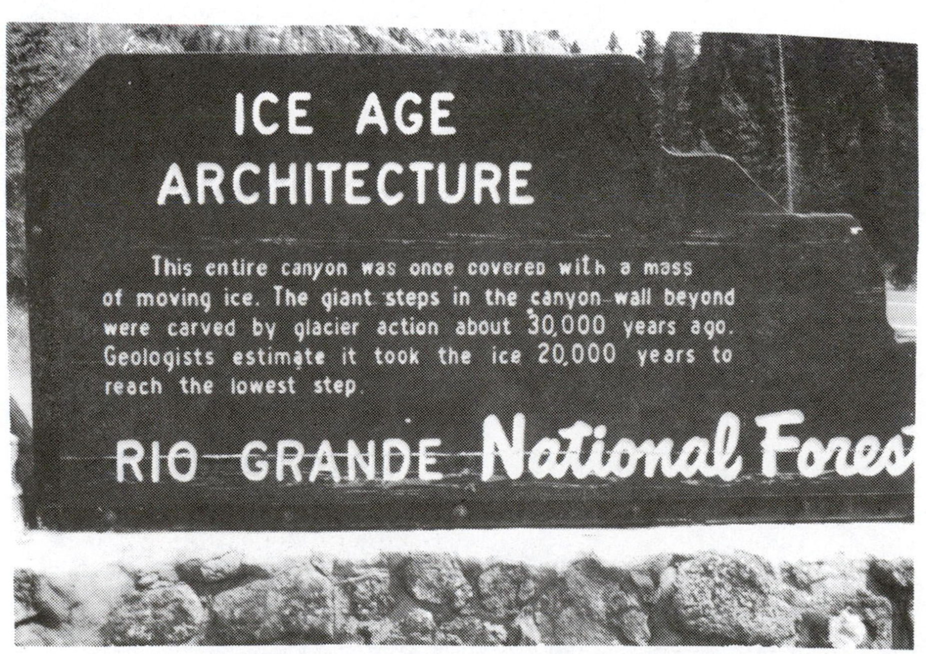

Fig. 298. A U.S. Forest Service sign calls attention to scour levels seen along the Rio Grande south of South Fork.

Fig. 299. A possible drumlin, sprinkled with glacial erratics, lies alone in a meadow near the San Juan Ranch in the headwaters of the Rio Grande. Ice moved from right to left.

Fig. 300. Just after Route 149 crosses Slumgullion Pass and heads down the Rio Grande drainage, South Clear Creek Falls is seen where the stream plunges over a hard flow of columnar basalt.

Fig. 301. Wagon Wheel Gap is a narrow canyon carved by the Rio Grande along Route 149 near Creede.

Fig. 302. A lonely mountain trail about 15 km. northeast of Creede leads to Colorado's secret National Monument. Wheeler Geologic Area was a monument, but it is so remote, near timberline at the east end of the San Juans, that even the U.S. Forest Service has trouble managing the site. The little patch of volcanic ash, of late Tertiary age, has very remarkable weathering features.

Fig. 303. Wheeler Geologic Area is strange enough for National Monument status but it takes a 16 km. dirt road and a 12 km. trail to reach it.

Fig. 304. Rain erosion eats away at Wheeler Geologic Area's ash, much as it would salt (but a little slower).

Fig. 305. Huge old mining buildings (arrows) are squeezed into a fault (?) notch near Creede, one of the most famous and productive districts in the state.

Fig. 306. Fishermen contemplate their chances at Rio Grande Reservoir, near the headwaters of the stream. They stand on glacially-scoured volcanic breccia, and the lake fills a classic, U-shaped trough. Weminuche Wilderness Area is on the far, south bank, and an excellent dirt road gives access on the north. The terrain is all volcanics, and the dam for this 8 km. long lake is a small man-made spillway built on a moraine that has been fortified by a massive rockslide of bus-sized boulders.

Fig. 307. Cochetopa Dome is BIG! Rising inside a moat-like basin twice as big as Crater Lake, Oregon, this boil of igneous rock rises 640 meters (2100 ft.) above the floor of the caldera, and is 8 km. wide. The moat is another 8 km. wide, making the whole thing 24 km. across, plus the slope beyond the moat. It is on Route 114, near the Continental Divide between Gunnison and Saguache.

The Gunnison Basin and The Black Canyon

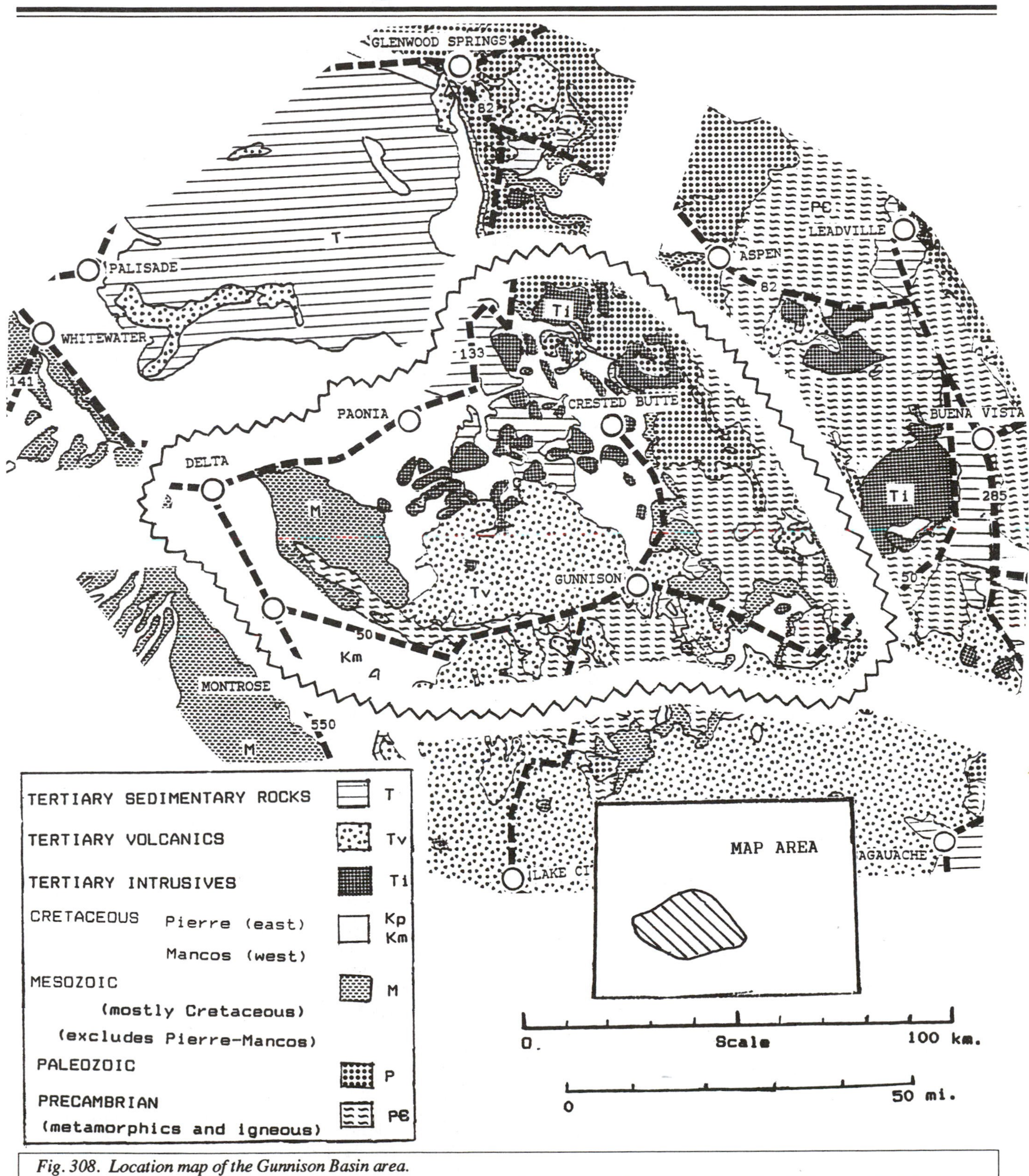

Fig. 308. Location map of the Gunnison Basin area.

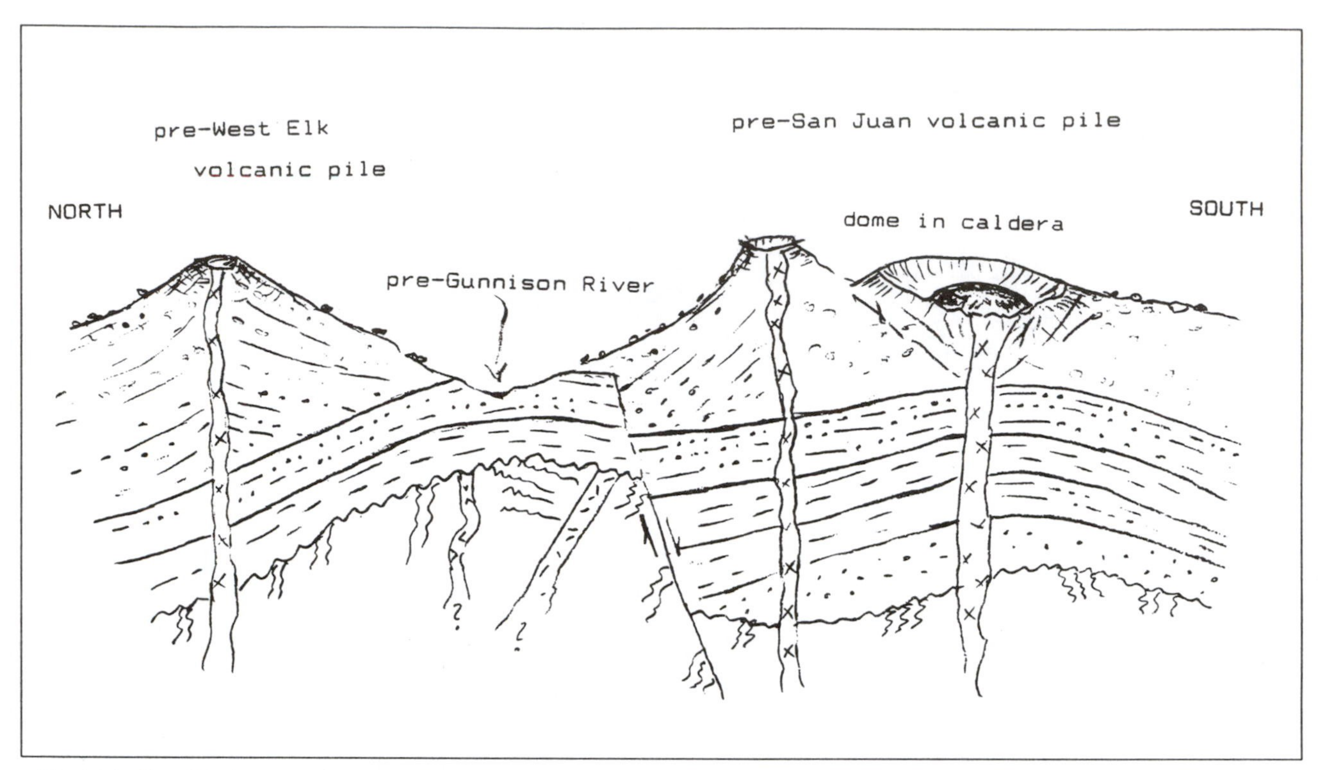

Fig. 309. Simplified crossection of the Black Canyon area about 20 million years ago. Volcanoes north and south have forced the ancestral Gunnison River to flow between the two volcanic piles.

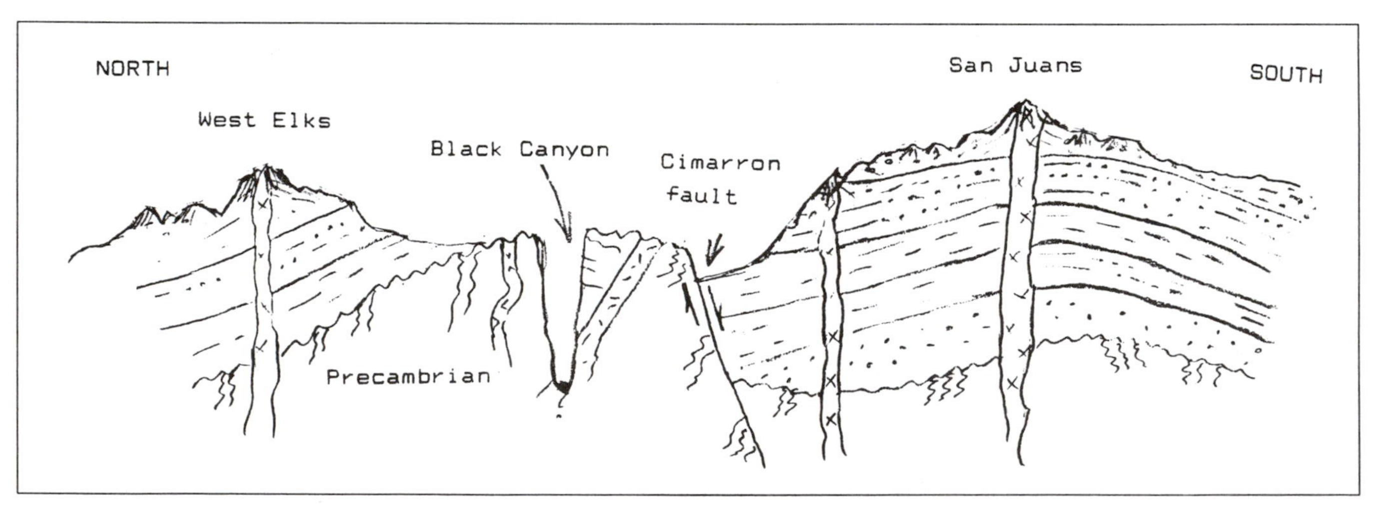

Fig. 310. Simplified crossection of the Black Canyon today. Note how the course of the river ended up slashing the very top of Black Ridge.

Fig. 311. Looking west at the Cimarron Fault near the small town of Sapinero on U.S. 50. Mancos shales are down-dropped against Precambrian metamorphics. The Gunnison River was trapped in a route down the axis of the Black Ridge, shown here on the right side of the photo. If a stream could pick a drainage route today, it would avoid the hard ridge, and find a route through the soft Mancos, where the highway is today.

Simply put, the Black Canyon of the Gunnison is a deep, narrow canyon, cut into the Precambrian metamorphics by the Gunnison River. But, that simple description is like saying Miss Universe is 59 kg. of water and a small percent of other minerals, standing almost two meters tall and moves. The canyon becomes spectacular by a strange coincidence of buried geology and a pair of volcanic piles. When the San Juan volcanic pile was popping away, there was another pile building on the north side of the pre-Gunnison River drainage. Some of the time the volcanic debris got ahead of the river and, no doubt, occasionally a large lake filled part of the Gunnison Basin.

When the volcanoes finally stopped spouting, the river began the horrendous task of removing all the sediment and other rocks in the area. Figures 309 and 310 show the before and after stages of erosion of the area. At first, in Fig. 309, the river is above a great thickness of volcanic material and a peculiar problem at depth. In Fig. 310 we see the river today. After it removed all the soft stuff it found itself on top of the very hard Precambrian rocks. The stream was trapped in its own canyon and had no choice but to continue downcutting in a hard rock that did not cave in on the sides. With so little side-sluffing, the canyon has nearly vertical walls. If there had been no uplift in the area to give the stream a steep gradient—thereby giving it the ability to downcut effectively, the drainage may have spilled

Fig. 312. A view from the north side of the river near Sapinero, showing Morrow Point Reservoir. U.S. 50 is in the distance. Black Ridge is the block on the right, and the Cimarron Fault is on the back side of the ridge. The river does not go through the gap in the center of the photo, but continues in the direction of the arrow, through the ridge.

Fig. 313. A Park Service boat on Morrow Point Reservoir, in the Black Canyon. Access to the lake is via a steep trail below the Blue Mesa Dam. The trail follows the old D&RG railroad grade. The rail grade then stays on the bottom of the lake, emerging again downstream from Morrow Point Dam. The Black Canyon National Monument is downstream from the dam and the dam is part of Curecanti National Recreation Area.

out of the basin in some other direction. So, the river was able to keep up with the uplift. Now the canyon goes down the axis of the highest terrain between the two volcanic piles. Note the thousands of meters of volcanic material that have been removed from the West Elks on the north and the San Juans in the south.

The location map (Fig. 308) shows a strip of Precambrian rock exposed most of the distance from the town of Gunnison to the town of Delta. The Gunnison River has uncovered that strip of Precambrian rock, which includes the Black Canyon northeast of Gunnison. The Gunnison drainage basin includes a large basin that includes Crested Butte. The North Fork of the Gunnison flows from the Paonia area and meets the main stream near Delta. The little patches of intrusive rock are mostly isolated laccoliths, giving the Gunnison area the nickname "laccolith capital of the world." Many laccoliths in the area are too small to be plotted on this scale of map.

Fig. 314. A boulder of Precambrian gneiss near the boat dock on Morrow Point Reservior. The official name of this formation is Black Canyon Schist, but much of the rock is banded, with poor cleavage-- a true gneiss.

Fig. 315. Curecanti Needle, on the right (south) side of Morrow Point Reservoir is the design image for the logo of the Denver and Rio Grande Railroad. The train once chugged around the base of this spire of Precambrian metamorphics, but now the railroad bed is under a hundred meters of water. The needle is much longer when the lake is empty.

Fig. 316. The Morrow Point Reservoir boat passes under Chipeta Falls, which spill over a horizontally-jointed granite intrusion. The rock looks like it is stratified, but spheroidal weathering gives a strong hint of the intrusive character of the rock.

Fig. 317. Colorado's highest cliff is the north wall of Black Canyon, viewed here from the south side. The 690 m. vertical cliff shows a light-colored network of Precambrian pegmatite dikes that riddled the gneiss and schists of the Black Canyon Formation.

Fig. 318. A dark line outlines Needle rock, a volcanic neck at Crawford Reservoir, just north of the Black Canyon. Tertiary stocks (laccoliths?) of the West Elk Mountains are in the background. This winter scene gives a hint to how severe winters can be in Colorado's high country.

Fig. 319. Along U.S. 50 between Montrose and Black Canyon is a grand display of slumps where the steep Mancos hillside gets too much irrigation water and shade. The slope is too steep, and fails frequently. The road and a railroad once traveled that side of the valley, but now, the dry side is used for U.S. 50, and the railroad is gone.

Fig. 320. Just downstream from the Black Canyon, the Gunnison River emerges into some modest, colorful Mesozoic canyons. In this view, the North Fork enters from the right just before the river cuts to the left and out of view. Placer gold in the Pleistocene terrace gravels, 20 m. and more above the present stream, are the object of this "diggins." The gold comes from a dozen or more tributaries to the Gunnison River.

Fig. 321. The town of Paonia is at the head of the North Fork, and these hills north of Paonia include the Mesaverde with its good bituminous coal. Westmoreland Coal Company works this major operation.

Fig. 322. Upstream from Paonia, one branch of North Fork is Muddy Creek, and it gets muddy from the soft Wasatch redbeds of Tertiary age and the shaly units of the Mesaverde. In 1986 about 10 km. of the east bank of Muddy Creek moved toward the stream. Frantic efforts kept the channel open, mostly, preventing a huge lake to form that would quickly flush out the dam and flood Paonia Reservoir and all towns downstream. This photo shows some of those cubic kilometers of moving hillside.

Fig. 323. A closer view of the 1986 slide on Muddy Creek. for about a week the trees groaned and popped as the slide continued, moving several meters per day. Ranches, roads, fences, ditches, springs and pipelines suffered.

Fig. 324. Upstream from Black Canyon, the main Gunnison has a string of man-made reservoirs: Crystal (6km. long), Morrow Point (15 km.) and Blue Mesa (26km.). This view is on the Lake Fork arm of Blue Mesa Reservoir, showing Precambrian metamorphics which are exposed along much of the reservoir.

Fig. 325. Looking north, across Blue Mesa Reservoir, the West Elk Breccia forms these pinnacles at Dillon Gulch. The tiny dark knob (arrow) is shown in Fig. 326, interpreted by the Park Service. This rock is a LAHAR deposit, or mudslide, rather than the glowing avalanche deposit that makes a welded tuff. Welded tuff is harder, and forms the darker cap on this exposure. Colorful beds of the Morrison are exposed below the cliffs.

Fig. 326. A Park Service interpretive display explains the Dillon Pinnacles. About 20 meters of this breccia is found on the south side of the river at the first good view of Blue Mesa Reservoir for east-bound travelers on U.S. 50. If the river was anywhere near its present course, the breccia dammed the river, forming a lake (a huge lake?).

Fig 327. A close view of the West Elk Breccia. Before this material was cemented, any sort of dam would be washed away easily--with perhaps spectacular flooding, possibly 20 millon years ago.

Fig 328. This thick, resistant, welded tuff holds up a broad, tree-covered area, south of the Gunnison, that provides the name "Blue Mesa." The Blue Mesa Reservoir is the largest lake in Colorado.

Fig 329. The West Elk Breccia appears again in this view at the head of Ohio Creek to form this cluster of spines called "the Castles."

Fig. 330. More West Elk Breccia is exposed overlooking the town of Gunnison, which is just out of this photo on the left.

Fig. 331. From Crested Butte to Paonia a secondary road climbs up Coal Creek to Kebler Pass and descends Anthracite Creek. This is a view of the "Great Dyke," which, of course, is a dike. The view is from half way up the north side of the Anthracite Range (a laccolith). This dike is one of several dikes that shoot northward out of the Anthracite laccolith. This one expands again to become the backbone of the Ruby Range. Mineralized redbeds of the Wasatch Formation, and other rocks, give the Rubies a purple cast, and hence the name.

Fig. 332. Crested Butte is wasted on skiers that plummet down the left (northwest) side shown here. With a heavy snowpack, the most numerous users of the area see only a blanket of white that hides the real geology. And, they cannot use Schofield Pass and the other scenic byways except with a snowmobile. The dozens of laccoliths that poke up in the Gunnison Basin penetrate the weak Mancos, and often warp the more competent Mesaverde into a crude anticline. Most of the Mesaverde has been removed by erosion since the middle Tertiary.

Fig. 333. Is it possible that Crested Butte is the model for the Geologic T-shirt sported by faculty and geology students at Western State College? The school in Gunnison is in the heart of some spectacular geology, but summers only seem to last from one to 15 days. Elevation is almost 2,400 m. (8,000 ft.).

Fig 334. Carbon Mountain, a laccolith, gets its name from coal mines around the base. Nearby intrusions tend to roast away some of the volatiles in coal, which change the economic "rank" of bituminous coal and can even metamorphose the coal to make anthracite coal. Most of the nation's anthracite coal is in the intensely-folded Appalachians in eastern Pennsylvania, where bituminous coals of Pennsylvanian age have been squeezed and cooked by metamorphism on a regional scale. Colorado's anthracite comes from the more selective "contact metamorphism" adjacent to an igneous source.

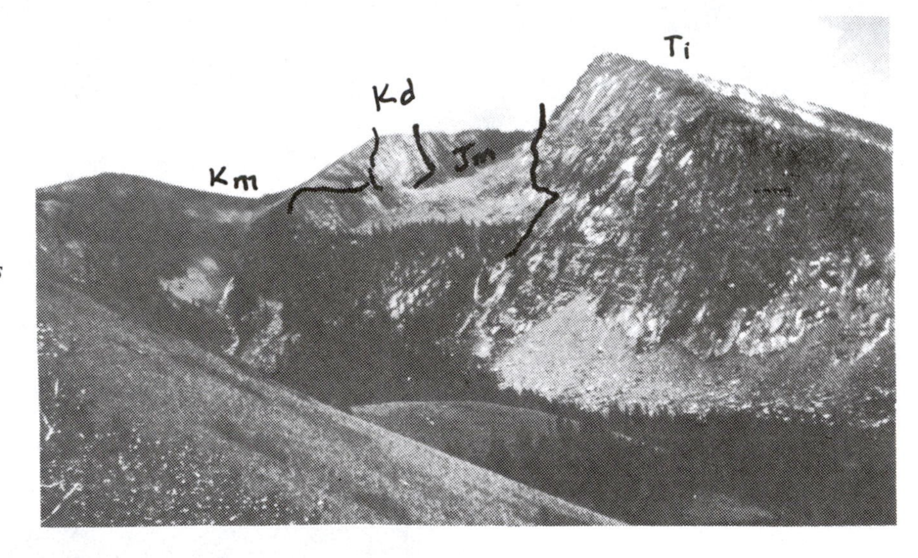

Fig. 335. Chair Mountain hangs over McClure Pass on the back road to Paonia from Glenwood Springs. This intrusion pushed the Dakota aside and roasted some of the nearby coal deposits in the Mesaverde. The author helped map and sample in this area in 1956 for Columbia Iron Mines, in search of a blend of coal for steel-making at the Geneva Steel plant in central Utah.

Fig. 336. Not all igneous activity near Gunnison made a laccolith. Here, a Tertiary dike (right) has cut sequences of Mesozoic sediments (left) as exposed in a roadcut east of Gunnison on U. S. 50.

Fig. 337. To keep the visitor from believing the Gunnison area has only simple laccoliths pushing up through the Mancos, the dam on Taylor Park Reservoir reminds us that some complicated fault troughs developed in the late Paleozoic. Here, the Cambrian Sawatch sands zig once and zag twice to show the power of some thrusting and folding of the area. The Precambrian metamorphics are part of the vertically-jointed material on the far right.

Fig. 338. Leaving the Gunnison basin north of Crested Butte the "minimum" road climbs over Schofield Pass, shown here. Hiding behind the skyline is the crest of the Elk Mountains, with their six fourteeners, including the two Maroon Bells. From Schofield Pass, one route down on the north side goes down Yule Creek, which passes the famed Yule Marble Quarry which provided the beautiful white stone for the Tomb of the Unknown Soldier at Arlington National Cemetery, Washington, D.C.

Fig. 339. Hartman Rocks is a field of unusual Precambrian granite south of Gunnison. Spheroidal weathering shouts "feldspars are weathering to clays!"

Fig. 340. At the Aberdeen Quarry, the granite has been exploited for building stones. The name comes from a nearly identical stone that has been quarried historically at Aberdeen, Scotland.

Fig. 341. This bathtub is one of the oddities in the Hartman Rocks.

Fig. 342. Tomichi Dome looms to the north as eastbound U.S. 50 exits the Gunnison Basin, about 30 km. east of Gunnison. This laccolith (?) pokes up through the Mancos next to Waunita Hot Springs, suggesting that, although these igneous features are long extinct, the subsurface in the area is hotter than most areas of the world. Tomichi resembles Cochetopa (Fig. 307), but is only about a third as large. Cochetopa is a dome rising from a collapsed caldera, and is surrounded by a distinctive, moat-like basin.

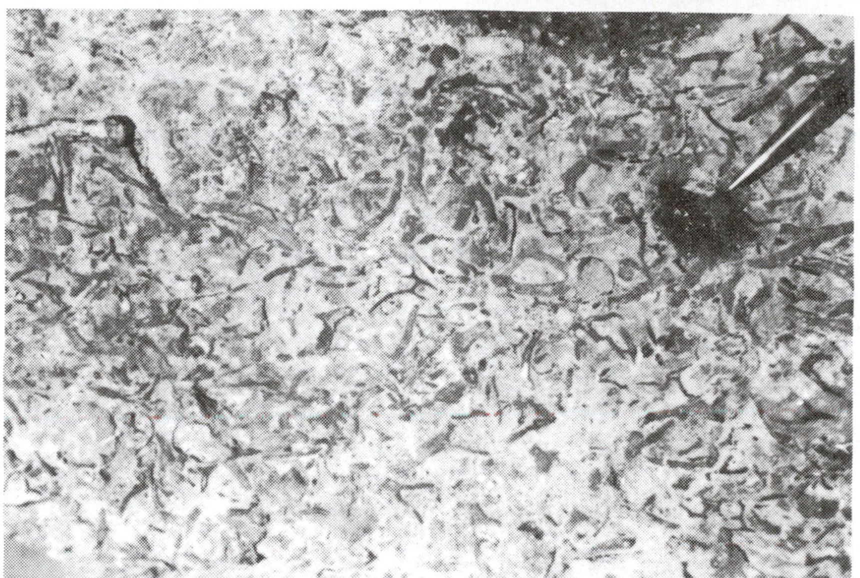

Fig. 343. Fossil bryozoans from Fossil Mountain. About 35 km. northeast of Gunnison is Fossil Mountain. Access requires an exhaustive hike, way above timberline, and most of the fossils are just like these in the photograph. Bryozoans are among the least interesting of the late Paleozoic fauna, even to paleontologists! Most people who attempt to climb to the fossils get side-tracked by the old gold camps along Quartz Creek and, especially, Gold Creek.

Fig. 344. This outcrop at Doyleville reminds us again what the sequence of sedimentary rocks is in the area. Surprise, a hogback of Dakota Sandstone! The Dakota pitches up abruptly at the Crookton fault, which marks the boundary between the Gunnison Basin (left) and the Sawatch uplift (right). From the Crookton fault, U.S. 50 climbs over Monarch Pass through Precambrian metamorphic rocks.

The Dinosaur Corner

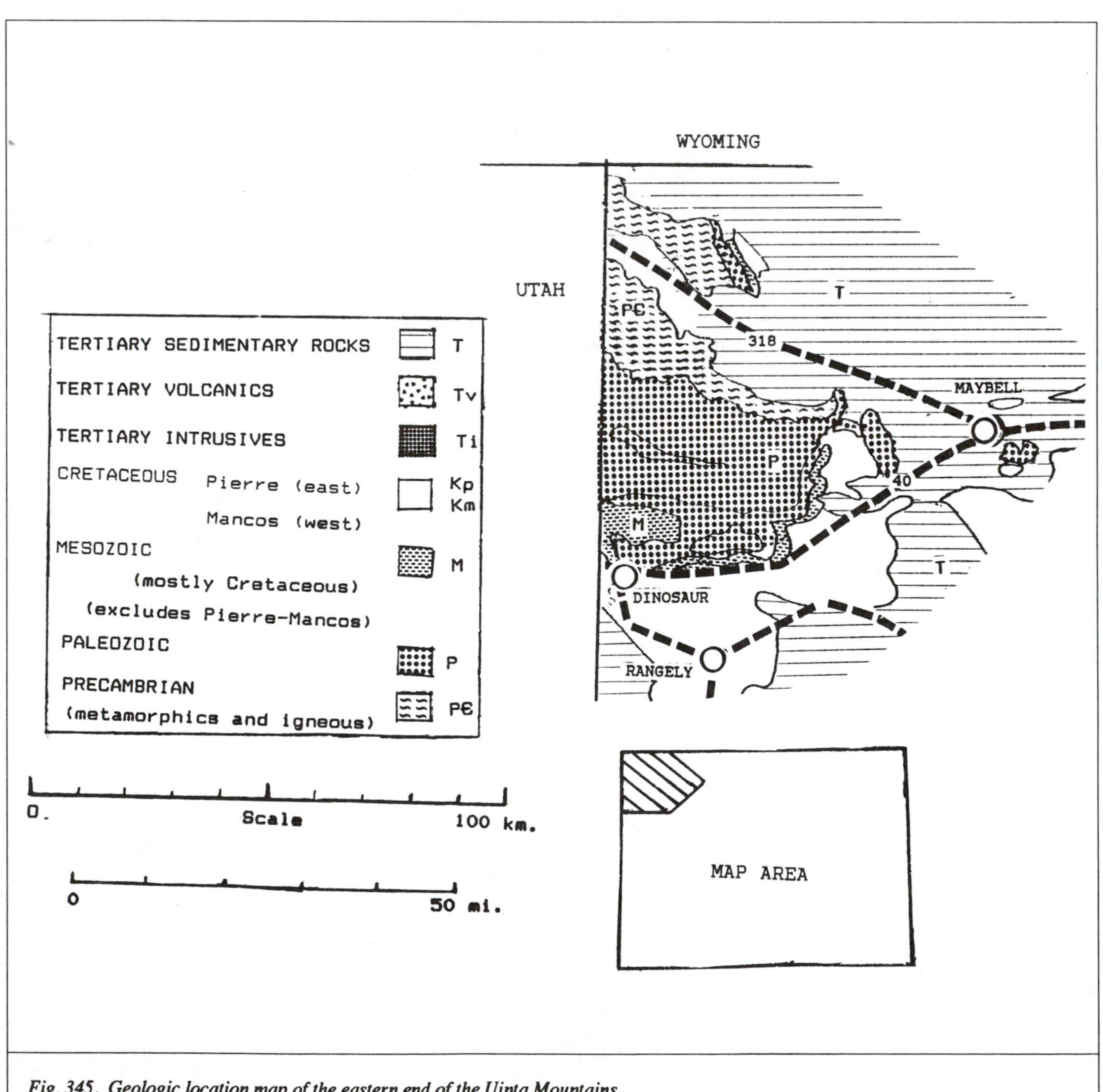

Fig. 345. Geologic location map of the eastern end of the Uinta Mountains.

The Uinta Mountains are an unusual lump in the central Rockies, as they extend east-to-west rather than north-south as is the trend of most ranges on the west side of the American Continents. Most of the range is in Utah, but the basic anticlinal structure extends at least 50 km. into Colorado, bringing a distinctive sequence of Precambrian sediments into view. Erosion has cut into the red sandstones and siltstones of the Uinta Mountain Group, exposing more than 1000 m. of the units at Colorado's Lodore Canyon on the Green River. The Precambrian is also exposed at Cross Mountain and Irish Canyon. The Uinta Mountain Group is genuine Precambrian age, and about 6 km. thick, but not strongly metamorphosed. The red sandstones are dense, but much of the sequence is silty and arkosic. Low grade metamorphism of the red shales produces some slate and phyllite, but not coarse schist.

The course of the Green River through the Uintas helps to decipher the Tertiary history of the general area. The Green River begins in the high, snowy peaks of the Wind River Mountains of Wyoming, then heads directly south towards Utah into the area of Flaming Gorge dam. Below the dam the river makes a sharp left turn (to the east) and runs about 10 km. into Colorado at Browns Park. Presumably the early Green River was blocked by an early stage of the Uintas. At the Gates of Lodore (the entrance to Lodore Canyon) the river turns sharply to the south again, cutting right through the Uinta anticline, and producing the colorful Lodore Canyon and Flaming Gorge.

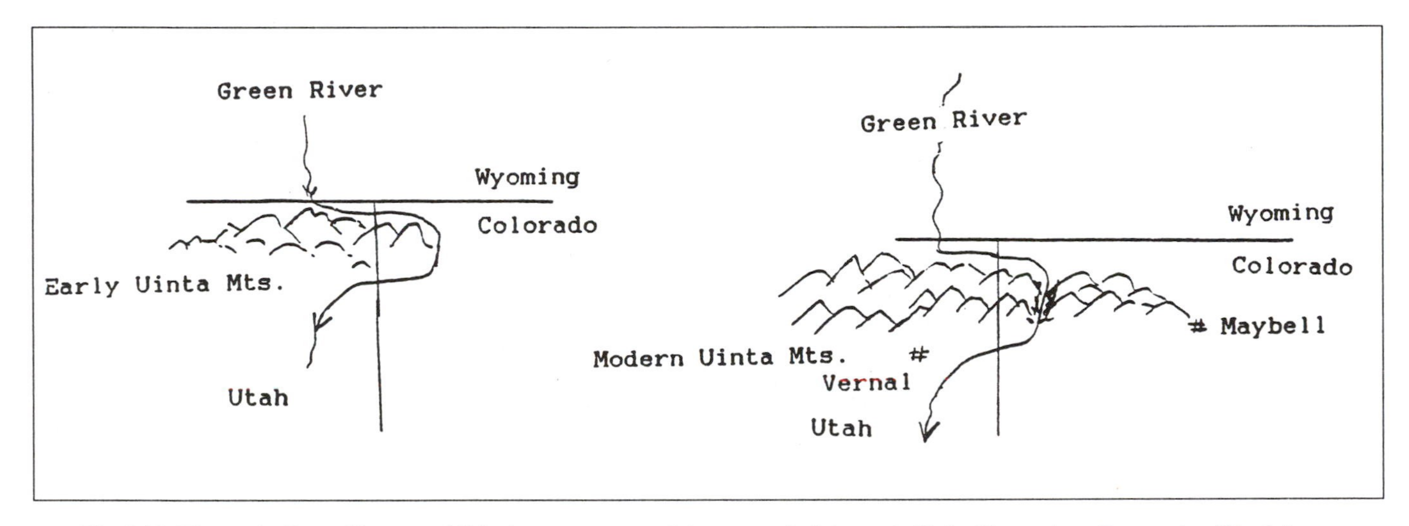

Fig. 346. The early Green River established a course around the east end of the early Unita Mountains. Renewed uplift of the region allowed the river to entrench itself in its channel. Continued erosion has exposed the sequence of hard rocks east of the river gorge that are really the east end of the Uinta Mountains.

Fig. 347. Echo Park surrounds Steamboat Rock in Dinosaur National Monument, just inside the state of Colorado. The Yampa River comes out of the shadow on the bottom right and joins the Green River, which then meanders out of the scene on the left. Much of the massive cliff is the clean Weber Sandstone of Pennsylvanian age.

Apparently the present path of the Green River was the eastern limit of the Uinta Mountains when it made the initial cut to the south. Soft, unconsolidated terrestrial sediments, or at most some soft, shaly Tertiary beds covered the area and the ancestral Green River picked an easy path around the eastern nose of the Uinta uplift. After the new path around the Uinta fold was established, a channel was scoured and the river became trapped in its new valley. The Uintas were uplifted along with the entire region, causing the Green River to rapidly cut down through the soft sediments and finally through thousands of meters of limestone and sandstones of the Mesozoic and Paleozoic formations. Eventually, a thousand meters of Precambrian were penetrated.

If a present-day stream were to leave the Wind River Mountains and head south, skirting the Uinta Mountains (without the Lodore Canyon route), a more likely path would be near Maybell, which is about 100 km. east of the present gorge. This is a case of an antecedent stream valley, in which the stream was in position and able to continue down-cutting, rather than be dammed off by the uplift. The stream was there before the obstacle.

There is a thin gravel unit, the Bishop Conglomerate, that flanked the Uintas in mid-Tertiary time. These gravels were stripped from the higher portions of the range, far to the west of the present Utah-Colorado boundary. At Harpers Corner, in Dinosaur National Monument, it is possible to stand on the Bishop Conglomerate and peer into the Green River Canyon to the river 700 m. below. The conglomerate under foot came from the source mountains on the opposite side of the canyon. It is certain that the Green River Canyon at Harpers Corner was not there when the Bishop Conglomerate was being distributed by streams that drained the Pre-Bishop Uinta Mountains.

The community of Dinosaur has capitalized on the current excitement about the huge lizards of the Morrison Formation. Back in the 1950's the author remembers working for the State Engineer's Office (Utah), when they were measuring its

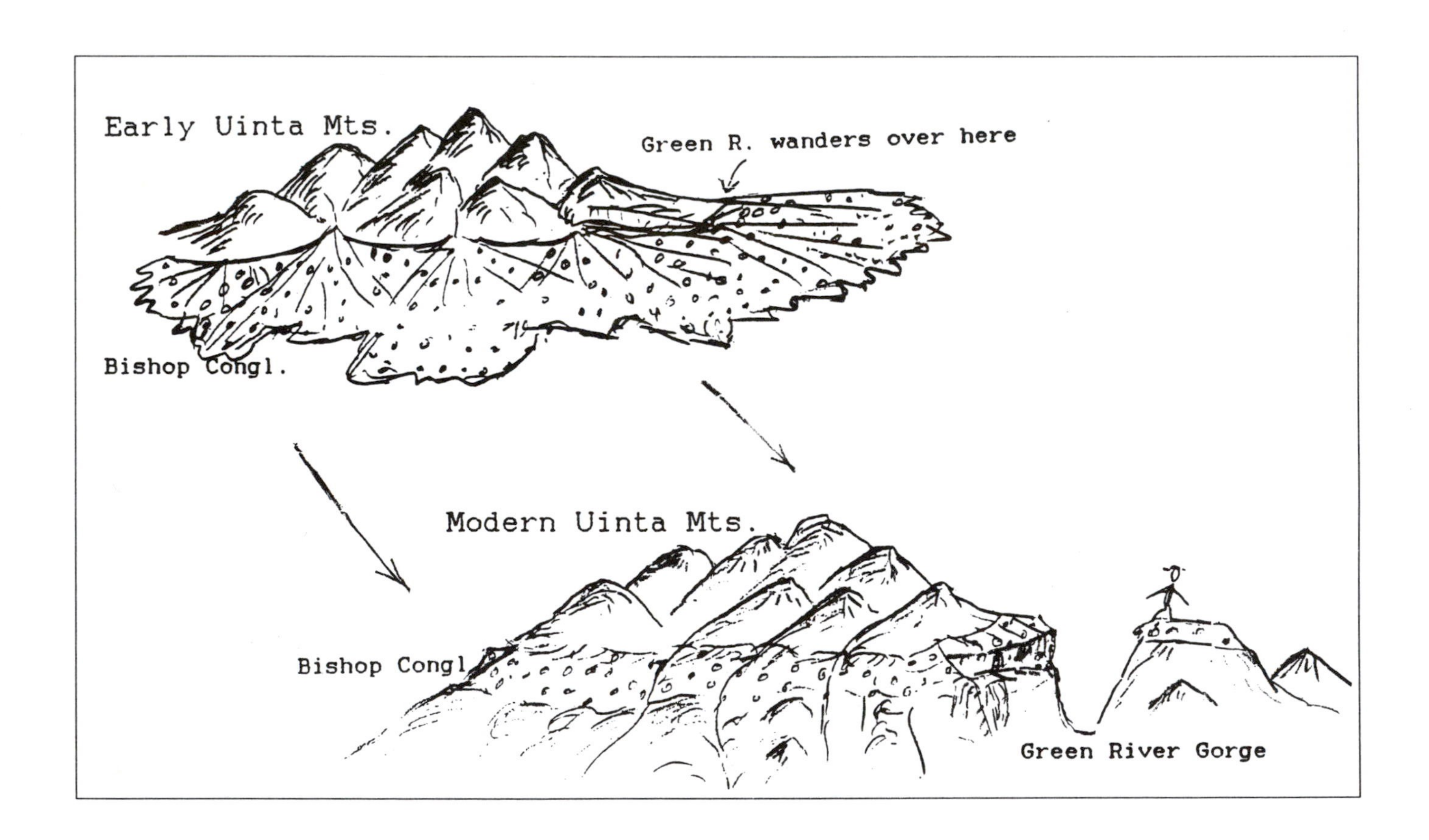

Fig. 348. During a long state of erosion, gravel was shed from the Uinta Mountains, forming an apron of coarse, rounded sediment named the Bishop Conglomerate. After an episode of uplift, the Green River became entrenched in its channel and was able to cut downward as fast as the mountains were uplifted. Outcrops of the Bishop Conglomerate are now found stranded east of the Green River gorge.

Fig. 349. Another air view of the Green River, carving its canyon in Dinosaur National Monument. The looping meanders suggest the river was running a sluggish, old age cycle before it was entrenched by uplift when the Uinta Mountains made their latest upsurge.

Fig. 350. The Weber Sandstone has sweeping (eolian?) crossbeds as seen from a boat at river level in Dinosaur National Monument. Note how high water erosion has etched the crossbedding in the very bottom of the photo.

water resources of the Uinta Basin. The community of Dinosaur, Colorado, was named Artesia in those days because the Dakota sands, and other aquifers, dip under the Uinta Basin, providing some limited artesian water. In those days there was also a difference between the liquor laws in the two states, and Artesia was the alcohol oasis for some of the people of eastern Utah. Tourists are usually impressed by the street signs in Dinosaur, which are written on dinosaur silhouettes unique for each street. There is Stegosaurus Road, Tyrannosaurus Street, and so on. However the idea has a strong negative effect because the signs are often stolen.

Oil and Coal are the big money-makers in the northwest corner of Colorado. Rangely field was developed in a giant, 32 km.-long textbook anticline which was even recognized by the sheep herders way back at the turn of the century. Oil seeps were found along the White River in "downtown" Rangely, so finding oil there was no surprise. The shallow discovery well was drilled in 1933, producing from fractured Mancos Shale. The remote location, however, forced the industry to wait until some deeper production was established in 1948 before building a pipeline to develop the field. Most of the production is from the deep Weber Sandstone of Pennsylvanian age. This sandstone is a clean, Utah version of the upper Maroon Formation. The field has produced nearly 800 million barrels of oil, making it a world-class giant oil field.

There is a little irony in the coal production in northwest Colorado. In the late 1970's the Deserado Coal Mine was developed just north of Rangely. The idea was to build a coal-fired generator at the mine, and send the electricity into the region's power grid from the point of origin. But there were some environmental problems permitting a coal plant in Colorado, so the company ended up sending the coal by conveyor belt five km. to the east where it could be cleanly loaded on an electric train (the only railroad in the Uinta Basin), and hauled back to the west (right past the mine) to Bonanza, UT, where the local authorities were delighted to have a new

Fig. 351. As viewed from Harpers Corner, right on the Colorado-Utah state line, the Mitten Park Fault shows drag-folding in the Pennsylvanian Morgan Formation which has slipped down against the Mississippian Madison (Leadville) Limestone, the white unit on the left, enhanced with some pencil lines. Some of the fingers of the "mittens" are highlighted to the right. For an observer on the river, these fingers are on the skyline.

Fig. 352. Rangely oil field is a real "giant," yet there is no crowding of the derricks or pumps. The field was developed after 1948 when efficient recoveries were programmed by having fewer wells, producing slower, so that the oil was more completely removed. The producer (Chevron) has completed a secondary water-flood, and is now in a carbon dioxide "tertiary recovery" cycle.

Fig. 353. A leaning street sign in the town of Rangely marks Camptosaurus Circle.

industry in a county with employment shortages. Bonanza is only eight kilometers inside Utah and the mine is 20 km. inside Colorado. By the time the coal is brought by conveyor to the load-out structure, it is 26 km. inside Colorado. It appears that the reason for not setting the powerplant in Colorado was because of pollution concerns, but imagine what direction the pollution goes from Bonanza when both states are in what is known as the "Prevailing Westerlies." Utah gets the benefit of the employment, much of the taxes, the sale of electricity to run the train, and can send the pollution plume back to Colorado. Deserado was asked once by an Illinois utility to send some coal to their plant for testing. There was not enough sulfur in the Colorado coal to use it in the eastern plant!

Fig. 354. The Deserado Coal mine runs coal out on the conveyor belt (right) and loads the coal on a private electric train for transport back past the mine to Utah where the coal is burned to produce electricity. The coal is from Cretaceous Mesaverde rocks.

Index

PUBLISHED AND DISTRIBUTED BY
YOUR GEOLOGIST
DELL R. FOUTZ
221 MESA AVENUE
GRAND JUNCTION, CO 81501
970/243-7088

ISBN 0-9640523-0-X